New Trends in Software Process Modeling

SERIES ON SOFTWARE ENGINEERING AND KNOWLEDGE ENGINEERING

Series on Software Engineering
and Knowledge Engineering Vol. 18

editors

Silvia T. Acuña
Universidad Autónoma de Madrid, Spain

Maria I. Sánchez-Segura
Universidad Carlos III de Madrid, Spain

New Trends in Software Process Modeling

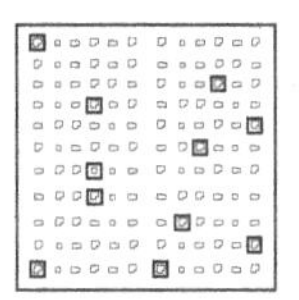

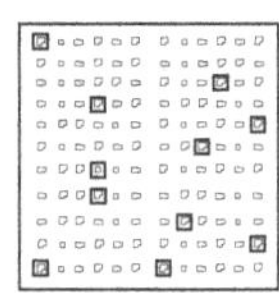

NEW JERSEY • LONDON • SINGAPORE • BEIJING • SHANGHAI • HONG KONG • TAIPEI • CHENNAI

Published by

World Scientific Publishing Co. Pte. Ltd.
5 Toh Tuck Link, Singapore 596224
USA office: 27 Warren Street, Suite 401-402, Hackensack, NJ 07601
UK office: 57 Shelton Street, Covent Garden, London WC2H 9HE

British Library Cataloguing-in-Publication Data
A catalogue record for this book is available from the British Library.

Series on Software Engineering and Knowledge Engineering Vol. 18
NEW TRENDS IN SOFTWARE PROCESS MODELING

ISBN-13 978-981-256-619-5
ISBN-10 981-256-619-8

Printed in Singapore

PREFACE

Silvia T. Acuña[1] and María I. Sánchez-Segura[2]

[1]*Departamento de Ingeniería Informática, Escuela Politécnica Superior, Universidad Autónoma de Madrid*
Avenida Tomás y Valiente 11, 28049 Madrid, Spain
E-mail: silvia.acunna@uam.es

[2]*Departamento de Informática, Universidad Carlos III de Madrid*
Avenida de la Universidad 30, 28911 Leganés, Madrid, Spain
E-mail: misanche@inf.uc3m.es

The software engineering discipline emerged in the 1960s as a consequence of the need to formalize the way software systems were developed. Since then a lot of research and development effort has gone into improving what have come to be termed software process models. The software process is a set of activities undertaken to manage, develop and maintain software systems, including the techniques to perform the tasks, the actors who execute the activities, their roles, constraints and the artifacts produced. Software process models are an abstraction of the software process which provides some useful information or insight for understanding, decision making and for enacting the actual process. Research in the 1990s was concerned with a variety of: not only prescriptive but also predictive and understanding-oriented[4] models. This was the consequence of a deeper understanding of the software process and the widening of the concerns of researchers who wished to investigate the impact of various organizational, social and economic factors on the software process.

One of the justifications for researching the software process is the view that the quality of the process has an impact on the quality of the software, and that process improvement positively influences the organization's performance. There are at least two reasons for building, evaluating and using process models[1]:

1. To achieve better ways of defining and guiding development, maintenance and evolution processes.
2. To achieve better ways of improving processes at the level of individual activities and the process as a whole.

Ever since the earliest days of software process research, the above two motivations have been at the heart of investigation carried out in this area[3]. There has, therefore, been significant progress in the above two directions, and hence they are not the focus of this book.

Software process modeling has recently been dealing increasingly with new challenges raised by the tests that the software industry has to stand such as, for example, the need to produce applications at Internet-time, pressures for budget cuts and customers who are demanding more complex software that is easier to use. This book is intended to help in the dissemination and understanding of new software process model trends. The new trends covered in this book are related to:

- Processes for open source software
- Software process simulation for process improvement, management and decision-support
- Peopleware[2], that is, the importance of people in the software process.

In other words, this book is intended to help readers understand the new software process models that are being developed to successfully manage new software development trends.

This book is structured as follows. The opening chapter explains an experience of implementing a process model for open source software. This is followed by three chapters (chapters 2, 3 and 4) that present the concept of the system dynamics approach to software processes improvement. Chapter 5 focuses on the new concept of people-oriented processes and what tools are available to support the enactment of these processes. Finally, chapter 6 recalls experience from describing the process model called E3 and the software system that supports this process model.

The discovery and building of process models for addressing new software development trends is known to be a long and costly process. Even so technological progress and the changing demands of today's society mean that the discovery and construction of new process models are always hot topics of research. One such new software development trend is the development of open source software. As such projects are a recent phenomena, the process models describing this type of development are not well known. The purpose of *chapter 1* then is to present a set of techniques for discovering, modeling, analyzing and simulating software development processes carried out in large open source software development projects based on public information accessible over the Web. Additionally, as an example of their applicability, the results of applying the defined techniques to a project with the above-mentioned characteristics, called NetBeans, are presented.

Simulation and dynamic modeling have been widely used as process improvement tools in industry. In recent years, this trend has also spread to the field of software process improvement. Hence, chapters 2, 3 and 4 focus on the description of work related to the use of simulation and dynamic modeling techniques in software processes.

Chapter 2 presents a process framework that combines traditional techniques with process modeling and simulation techniques that support a qualitative and quantitative evaluation of the development process. This evaluation can be used to improve the software process, and is also a decision-making aid. The use of this framework can help software organizations achieve a higher process capability following to SEI's CMM (Capability Maturity Model)[5].

Chapter 3 includes a survey of the main process simulation applications since the 1990s. Additionally, this chapter describes IMMoS (Integrated Measurement Modeling and Simulation)[6], a method for developing goal-oriented dynamic simulation models. By way of an illustration of the applicability of the IMMoS model, several cases of software process simulation models that were developed to support learning and decision making in software organizations within the automobile manufacturing industry are described.

Chapter 4 presents an approach based on high level modeling for software projects. This approach separates the description of a particular project from the knowledge embedded in a software project model. The aim is to make useful complex system dynamics-based models that are built and adapted not only by experts but also by practitioners. Along these lines, this chapter describes a modeling and simulation process for system dynamics that allows the development of domain models and their specialization for particular problems.

Chapter 5 addresses software development by means of people-oriented process models. These models have turned out to be very beneficial because they improve the quality of the interaction between people and processes. The chapter is divided into three parts focusing on the capture, visualization and use of the information by the people involved in the software development process. With respect to information capture, this chapter describes different knowledge process types and discusses the application of the GQM (Goal Question Metric) paradigm for data collection and/or to measure the particular process for which the data are captured. As regards the part of the process model related to the visualization of the information needed by each developer involved in a particular process, the generation of documents, role-based workspaces and control centers for software development are discussed. The use of the captured information is another important issue and is illustrated by discussing aspects concerning the management of previous experiences to assure that each experience can improve future development processes.

Chapter 6 provides input for readers interested in learning about the evolution of process models. This chapter examines the evolution of an existing process model (E3) and the software system that supports this model, called the E3 system. E3 is a process model conceived to provide help for process/project managers, who construct and maintain models, and for practitioners, who use software process models. The chapter is a *post-mortem* analysis of the decisions that led to the E3 system requirements definition and gives insight into what principles any process model should follow if it is to remain useful years after being initially conceived.

Acknowledgments

We would like to thank all the authors of the submitted chapters whose research has made this edited book possible. Particular thanks are due to Rachel Elliott who assembled the material and ensured that the presentation was consistent. We are especially grateful to Natalia Juristo and Juan F. Ramil for comments about the preface. We are also deeply indebted to Jens Heidrich, Chris Jensen, Cláudia Maria Lima Werner, Jürgen Münch, Márcio De Oliveira Barros, Rodion M. Podorozhny, Isabel Ramos, Günther Ruhe, Mercedes Ruiz and Guilherme Horta Travassos for helping us to improve the chapters of this book.

References

1. Acuña, S.T. and N. Juristo. *Software Process Modeling*, International Series in Software Engineering, Vol. 10, Springer, NY, 2005.
2. DeMarco, T. and T. Lister. *Peopleware: Productive Projects and Teams*, 2nd ed. Dorset House, NY, 1999.
3. Derniame, J.C., B.A. Kaba and D. Wastell (Eds.). *Software Process: Principles, Methodology and Technology*, Lecture Notes in Computer Science 1500, Springer, Berlin, 1999.
4. Kellner, M.I., R.J. Madachy and D.M. Raffo. Software Process Simulation Modeling: Why? What? How?, *Journal of Systems and Software*, 46, 2-3, 91-105, 1999.
5. Paulk, M.C., B. Curtis, M.B. Chrissis and C.V. Weber. The Capability Maturity Model for Software, Version 1.1, *IEEE Software*, 10, 4, 18-27, 1993.
6. Pfahl, D. and G. Ruhe. IMMoS: A Methodology for Integrated Measurement, Modeling, and Simulation, *Software Process Improvement and Practice*, 7, 189-210, 2002.

CONTENTS

Chapter 1

DISCOVERING, MODELING, AND RE-ENACTING OPEN SOURCE SOFTWARE DEVELOPMENT PROCESSES: A CASE STUDY

Chris Jensen and Walt Scacchi

Institute for Software Research
Donald Bren School of Information and Computer Science
University of California, Irvine
Irvine, CA USA 92697-3425
Email: {cjensen, wscacchi}@ics.uci.edu

Software process discovery has historically been a labor and time intensive task, either done through exhaustive empirical studies or in an automated fashion using techniques such as logging and analysis of command shell operations. While empirical studies have been fruitful, data collection has proven to be tedious and time consuming. Existing automated approaches have very detailed, low level but not rich results. We are interested in process discovery in large, globally distributed organizations such as the NetBeans open source software development community, which currently engages over twenty thousand developers distributed over several continents working collaboratively, sometimes across several stages of the software lifecycle in parallel. This presents a challenge for those who want to join the community and participate in, as well as for those who want to understand these processes. This chapter discusses our efforts to discover selected open source processes in the NetBeans community. We employ a number of data gathering techniques ranging from ethnographic to semi-structured to formal, computational models, which were fed back to the community for further evaluation. Along the way, we discuss collecting, analyzing, and modeling the data, as well as lessons learned from our experiences.

1. Introduction

The Peopleware vision is an attempt to provide insight into the social qualities of project management that may lead to project success or failure. In a similar sense, open source software development (OSSD) has been effective in providing online social workscapes that have become the focus of attention in industry and research conferences alike. However, in order to understand and participate in these processes, people new to these processes must first discover what they are and how they operate. The goal of our work is to develop new techniques for discovering, modeling, analyzing, and simulating software development processes based on information, artifacts, events, and contexts that can be observed through public information sources on the Web. Our problem domain examines processes in large, globally dispersed OSSD projects, such as those associated with the Mozilla Web browser, Apache Web server[1], and Java-based integrated development environments for creating Web applications like NetBeans[2] and Eclipse[3]. The challenge we face is similar to what prospective developers and corporate sponsors who want to join a given OSSD project face, and thus our efforts should yield practical results.

Process models are *prescriptive* if they state what activities should be done or *proscriptive* if they describe what activities could be done. With process discovery, our task is to create *descriptive* models by determining what activities have been done. OSSD projects, however, do not typically employ or provide explicit process model prescriptions, proscriptions, descriptions, or schemes other than what may be implicit in the use of certain OSSD tools for version control and source code compilation. In contrast, we seek to demonstrate the feasibility of automating the discovery of software process workflows in projects like NetBeans by computer-assisted search and analysis of the project's content, structure, update and usage patterns associated with their Web information spaces. These spaces include process enactment information such as informal task prescriptions, community and information structure and work roles, project and product development histories, electronic messages and communications patterns among project participants[4, 5, 6]. Similarly, events that denote updates to these sources are also publicly accessible, and thus suitable for analysis. Though traditional ethnographic approaches to software process discovery[7] net a wealth of information with which to model, simulate, and analyze OSSD processes, they are time and labor-intensive. As a result, they do not scale well to the study of multiple OSSD development projects of diverse types in a timely manner. Subsequently, this suggests the need for a more automated approach that can facilitate process discovery.

In our approach, we examine three types of information in the course of discovering and modeling OSSD processes. First are the kinds of OSSD artifacts (source code files, messages posted on public discussion forums, Web pages, etc.). Second are the artifact update events (version release announcements, Web page updates, message postings, etc.). Third are work contexts (roadmap for software version releases, Web site architecture, communications systems used for email, forums, instant messaging, etc.) that can be detected, observed, or extracted across the Web. Though such an approach clearly cannot observe the entire range of software development processes underway in an OSSD project (nor do we seek to observe or collect data on private communications), it does draw attention to what can be publicly observed, modeled, or re-enacted at a distance.

Our approach relies on use of a process meta-model to provide a reference model that associates these data with software processes and process models[8]. Whereas the meta-model describes the attributes of process events and how they may be arranged (i.e. the language of the process), the reference model describes types and known instances of those attributes. As such, we have been investigating what kinds of processing capabilities and tools can be applied to support the automated discovery and modeling of selected software processes (e.g., for daily software build and periodic release) that are common among many OSSD projects. The capabilities and tools include those for Internet-based event notification, Web-based data mining and knowledge discovery, and previous results from process discovery studies. However, in this study, we focus on identifying the foundations for discovering, modeling, and re-enacting OSSD processes that can be found in a large, global OSSD project using a variety of techniques and tools.

2. Related Work

Process event notification systems have been used in many contexts, including process discovery and analysis[9, 10]. However, of the systems promising automated event notification, many require process performers to obtain, install, and use event monitoring applications on their own machines to detect when events occur. While yielding mildly fruitful results, this approach is undesirable for several reasons, including the need to install and integrate remote data collection mechanisms with local software development tools.

Prior work in process event notification has also been focused on information collected from command shell histories, applying inference techniques to construct process model fragments from event patterns[11]. They advise that rather than seeking to discover the entire development process, to instead focus on creating partial process specifications that may overlap with one another. This also reflects variability in software process enactment across iterations. This imparts additional inconvenience on the user and relies on her/his willingness to use the particular tools that monitor and analyze command shell events. By doing so, the number of process performers for whom data is collected may be reduced well below the number of participants in the project due to privacy concerns and the hassles of becoming involved. While closed source software engineering organizations may mediate this challenge by leveraging company policies, OSSD projects lack the ability to enforce or the interest to adopt such event capture technology.

Recently, there have been a number of developments focused on mining software repositories[12, 13]. While these have yielded interesting insights into patterns of software development in OSSD communities, most of the work has focused on low-level social network analysis of artifacts and agents of software development rather than processes of software development.

Lastly, while process research has yielded many alternative views of software process models, none has yet been proven decisive or clearly superior. Nonetheless, contemporary research in software process technology, such as Lil Jil process programming language[14, 15] and the PML process modeling and enactment language[16], argues for analytical, visual, navigational and enactable representations of software processes. Subsequently, we find it fruitful to convey our findings about software processes, and the contexts in which they occur, using a mix of both informal and formal representations of these kinds. Thus, we employ this practice here.

3. Problem Domain

We are interested in discovering, modeling, and simulating re-enactment of software development processes in large, Web-based OSSD projects. Such projects are often globally distributed efforts sometimes involving hundreds or thousands of developers collaborating on products constituting thousands to millions of source lines of code without meeting face-to-face, and often without performing modern methods for software engineering[5]. Past approaches have shown process discovery to be difficult, yielding limited results. However, the discovery methods we use are not random probes in the dark. Instead, we capitalize on contextual aids offered by the domain and captured in the process reference model. Some of these include:

- Web pages, including project status reports and task assignments
- Asynchronous communications among project participants posted in threaded email discussion lists
- Transcripts of synchronous communication via Internet chat
- Software problem/bug and issue reports
- Testing scripts and results
- Community newsletters
- Web accessible software product source code directories
- Software system builds (executable binaries) and distribution packages
- OSS development tools in use in an OSSD project
- OSS development resources, including other software development artifacts.

Each OSSD project has locally established methods of interaction and communication, whether explicit or implicit[5, 6]. These collaboration modes yield a high amount of empirically observable process evidence, as well as a large degree of unrelated data. However, information spaces are also dynamic. New artifacts are added, while existing ones are updated, removed, renamed and relocated, else left to become outdated. Artifact or object contents change, and project Web sites get restructured. In order to capture the history of process evolution, these changes need to be made persistent and shared with new OSSD project members. While code repositories and project email discussion archives have achieved widespread use, it is less common for other artifacts, such as instant messaging and chat transcripts, to be archived in a publicly available venue. Nonetheless, when discovering a process in progress, changes can de detected through comparison of artifacts at different time slices during the development lifecycle. At times, the detail of the changes is beneficial, and at other times, simply knowing what has changed and when is all that is important to determining the order (or control flow sequence) of process events or activity. To be successful, tools for process discovery must be able to efficiently access, collect, and analyze the data across the project Web space. Such data includes public email/mailing list message boards, Web page updates, notifications of software builds/releases, and software bug archives in terms of changes to the OSS information space[5, 6].

How the project organizes its information space may indicate what types of artifacts they generate. For example, a project Web page containing a public file directory named "x-test-results" can be examined to determine whether there is evidence that some sort of testing (including references to test cases and test results) has been conducted. Furthermore, timestamps associated with file, object, or Web page updates provide a sense of recent activity and information sharing. Similarly, when new branches in the Web site are added, we may be able to detect changes in the process or discover previously unknown activities. Elsewhere, the types of artifacts available on the site can provide insight into the project development process. Further investigation may excavate a file named "qa-functional-full" under the "x-test-results" directory, indicating that that functional testing has been performed on the entire system. Likewise, given a graphic image file (a Web-compatible image map) and its name or location within the site structure, we may be able to determine that an image named "roadmap2003" may show the progression that the project has made through the year of 2003, as well as future development milestones. This process "footprint" tells us that the some informal planning has been done. In some cases, artifacts containing explicit process fragments have been discovered, which may then be validated against the discovered process to determine whether the project is enacting the process as described. Whereas structure and content can tell us what types of activities have been performed, monitoring interaction patterns can tell us how often they are performed and what activities the project views as more essential to development and which are peripheral.

4. Field Site and Process Description

To demonstrate the viability of our process discovery approach, we describe how we apply it through a case study. For this task, we examine a selected process in the NetBeans[2] OSSD project. The NetBeans project started in 1996 as a student project before being picked up and subsequently made an OSSD project by Sun Microsystems. The NetBeans project community is now an effort combining dozens of organizations (4 distributing releases, 42 writing extensions, and 21 building tools based on the NetBeans platform)[1] and boasts of over one hundred thousand developers around the globe[2]. The scale of the project thus necessitates developers to transparently coordinate their efforts and results in a manner that can be accessed and persist on the community Web site. As demonstrated in the previous section, this coordination evidence forms the basis from which processes may be identified and observed.

The *requirements assertion and release* process was chosen for study because its activities have short duration, are frequently enacted, and have a propensity for available evidence that could potentially be extracted using automated technologies. The process was discovered, modeled informally and formally, then prototyped for analysis and re-

---

[1] http://www.netbeans.org/about/third-party.html, as of October 2004

[2] http://www.netbeans.org/community/news/index.html#494, as of October 2004

enactment. The next two sections describe the methods we use to discover, model, and re-enact the requirements and release process found in the NetBeans OSSD project. Along the way, we present descriptions of the process under study using informal, semi-structured, and formal models, and the formal models are then further analyzed through a process enactment simulator we use for process re-enactment. Process re-enactment in turn allows us to further validate our models, as well as serve to refine and improve the discovered processes as feedback to the OSSD project in the study on possible process improvement opportunities.

5. Process Discovery and Modeling

The discovery and modeling approach used in this case study consisted of defining the process meta-model and reference model for the selected process, as described above. Next, we gathered data from the community Web, indexing it according to the reference model with an off-the-shelf search engine and correlating it based on its structure, content, and update context (usage data being unavailable for this project). Our experience has shown that it is best to view process discovery and modeling as a progressive activity. That is, we utilize several models at different levels of abstraction that become progressively more formal. The first of these depicts activity scenarios that reflect the correlation of tools, resources, and agents in the performance of development activities (i.e. instances of the attributes of the process meta-model). These are then refined into a flow graph illustrating more concretely, the order in which the activity scenarios are performed and lastly, a formal, computational process model. Moreover, progressive discovery can reduce collection of unrelated "noisy" process data by using coarsely grained data to direct the focus of discovery of more finely grained data.

As our results stem from empirical observations of publicly available artifacts of the NetBeans community Web, they face certain threats to validity. Cook *et al.*[17] demonstrated the validity of using the kinds of observations described above in terms of constructing process models, as well as showing internal and external consistency. Unlike Cook and Wolf, we apply *a priori* knowledge of software development to discovering processes. Instead of randomly probing the information space, we use the reference model to help locate and identify possible indicators that a given activity has occurred. This reference model was devised through a review of several open source communities (NetBeans, Mozilla, and Apache, among others). However, it must be updated to reflect the evolution of the types (and names) of tools, activities, roles, and resources in OSSD, in particular those of the particular community subject to process discovery to maintain the validity of our methodology. The full results of our case study may be found in[18]. Subsequent discovery of our study by the community and our discussions with prominent community members that followed verified our results and provided additional insight. This feedback allows both process modelers and process participants opportunities for mutual improvement of methods and the outputs they produce (i.e. process modeling techniques and process models, as well as software development processes and software products). The discussion of our process discovery and modeling methods and results follows next.

The discovery of processes within a specific OSSD project begins with a cursory examination of the project Web space in order to ascertain what types of information are available and where that information might be located within the project Web. The information gathered here is used to configure the OSSD process reference model[19]. This framework provides a mapping between the tool, resource, activity, and role names discovered in the community Web with a classification scheme of known tools, resources, activities, and roles used in open source communities. This step is essential to permit association of terms such as "CVS" with source versioning systems, which have certain implications in the context of development processes. The project site map provided not only a breakdown of project Web pages within each section, but also a timestamp of the latest update. This timestamp provides empirical evidence gathered from project content that reflects the process as it is currently enacted, rather than what it has evolved from.

Guided by our framework, searching the "about" sub-section of the project Web site provided information on the NetBeans technologies under development, as well as the project structure (e.g., developer roles, key development tasks, designated file-sharing repositories, and file directories) and the nature of its open source status. This project structure is a basis for understanding current development practices. However, it also details ways for outsiders to become involved in development and the community at large[3]. The modes of contribution can be used to construct an initial set of activity scenarios, which can be described as *use cases* for project or process participation.

Though best known as a tenet of the unified modeling language (UML), use cases can serve as a notation to model scenarios of activities performed by actors in some role[20, 7]. The site map also shows a page dedicated to project governance hyperlinked three layers deep within the site. This page exposes the primary member types, their roles and responsibilities, which suggest additional use cases. Unlike those found through the modes of contribution, the project roles span the breadth of the process, though at a higher level of abstraction. Each use case can encode a process fragment. In collecting use cases, we can extract out concrete actions that can then be assembled into a process description to be modeled, simulated, and enacted.

When aggregated, these use cases can be coalesced into an informal model of a process and its context rendered as a *rich interactive hypermedia*, a semi-structured extension of Monk and Howard's[21] rich picture modeling construct. The rich hypermedia shown in Figure 1 identifies developer roles, tools, concerns, and artifacts of development and their interaction, which are hyperlinked (indicated as underlined phrases) to corresponding use cases and object/role descriptions (see Figure 2). Such an informal computational model can be useful for newcomers to the community looking to become involved in development and offers an overview of the process and its context in the project, while abstracting away the detail of its activities. The use cases also help identify the requirements for enacting or re-enacting the process as a basis for validating, adapting, or improving the process.

[3] http://www.netbeans.org/community/contribute, as of June 2004

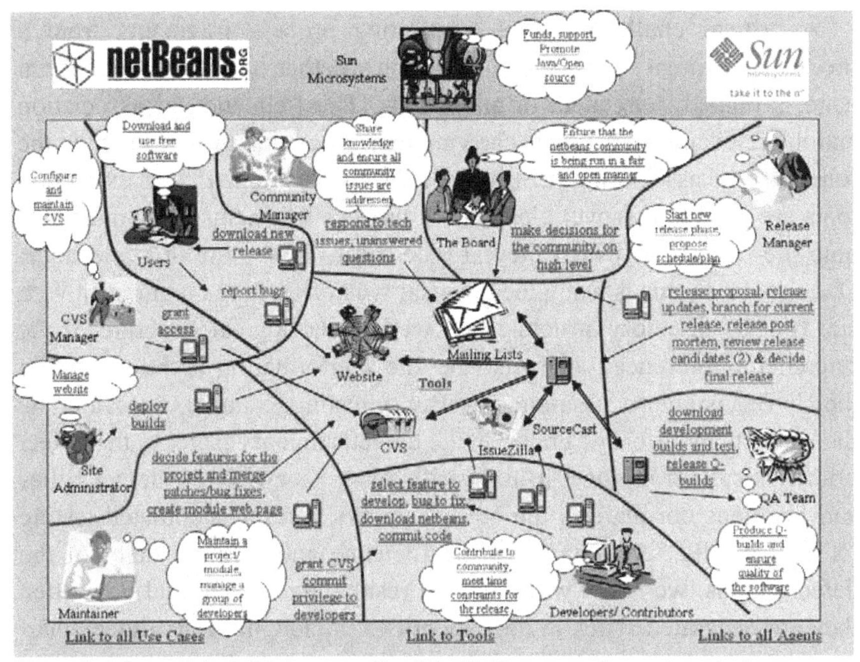

Figure 1. A hyperlinked rich hypermedia of the NetBeans requirements and release process[18]

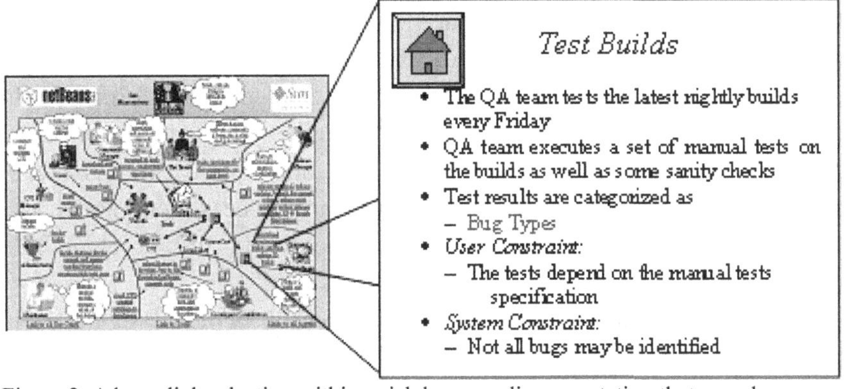

Figure 2. A hyperlink selection within a rich hypermedia presentation that reveals a corresponding use case

A critical challenge in reconstructing process fragments from a process enactment instance is in knowing whether or not the evidence at hand is related, unrelated, or anomalous. The frequency of association and the relevance of artifacts carrying the association may strengthen the reliability of associations constructed in this fashion. If text extraction tools are used to discover elements of process fragments, they must also note the context in which they are located to determine this relevance. One way to do this is using the physical structure of the community Web site (i.e. its directory structure), as well as the logical structure of the referencing/referenced artifacts (the site's information architecture). In the NetBeans quality assurance (Q-Build) testing example, we can relate the "defects by priority" graph on the defect summary page[4] to the defect priority results from the Q-Build verification. Likewise, the defect tallies and locations correlate to the error summaries in the automated testing (XTest) results[5]. By looking at the filename and creation dates of the defect graphs, we know which sets of results are charted and how often they are generated. This, in turn, identifies the length of the defect chart generation process, and how often it is executed. The granularity of process discovered can be tuned by adjusting the search depth and the degree of inference to apply to the data gathered. An informal visual representation of the artifacts that flow through the requirements and release process is shown in Figure 3.

These process fragments can now be assembled into a formal PML description of the selected processes[16]. Constructing such a process model is facilitated and guided by use of an explicit process meta-model[8]. Using the PML grammar and software process meta-model, we created an ontology for process description with the Protégé-2000 modeling tool[22].

The PML model builds from the use cases depicted in the rich hypermedia, then distills from them a set of actions or sub-processes that comprise the process with its corresponding actor roles, tools, and resources and the flow sequence in which they occur. A sample result of this appears in Figure 4.

---

[4] http://qa.netbeans.org/bugzilla/graphs/summary.html as of March 2004
[5] http://www.netbeans.org/download/xtest-results/index.html as or March 2004

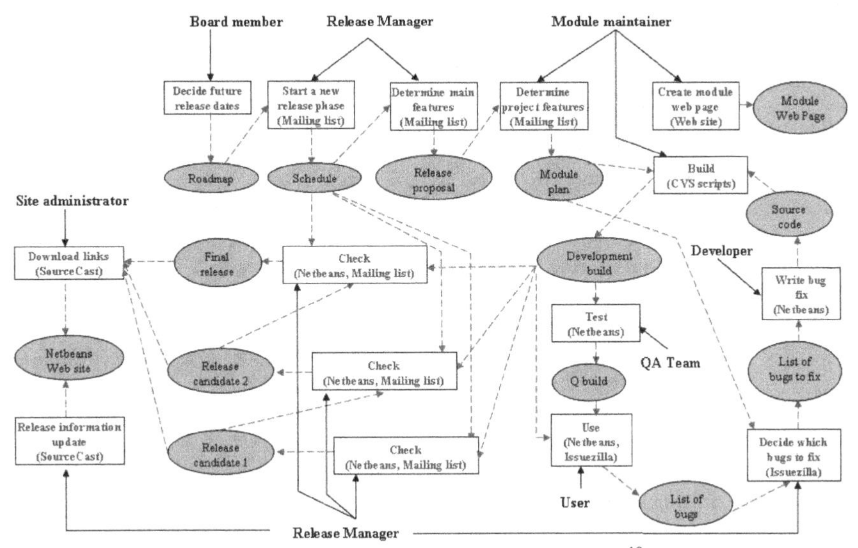

Figure 3. NetBeans Requirements and Release process flow graph[18]

6. Process Re-enactment for Deployment, Validation, and Improvement

Since their success relies heavily on broad, open-ended participation, OSSD projects often have informal descriptions of ways members can participate, as well as offering prescriptions for community building[5]. Although automatically recognizing and modeling process enactment guidelines or policies from such prescriptions may seem a holy grail of sorts for process discovery, there is no assurance that they accurately reflect the process as it is enacted. However, taken with the discovered process, such prescriptions begin to make it possible to perform basic process validation and conformance analysis by reconciling developer roles, affected artifacts, and tools being used within and across modeled processes or process fragments[23].

```
1.  sequence Test {
2.    action Execute automatic test scripts {
3.      requires { Test scripts, release binaries }
4.      provides { Test results }
5.      tool { Automated test suite (xtest, others) }
6.      agent { Sun ONE Studio QA team }
7.    }
8.    action Execute manual test scripts {
9.      requires { Release binaries }
10.     provides { Test results }
11.     tool { NetBeans IDE }
12.     agent { users, developers, Sun ONE Studio QA team,
          Sun ONE Studio developers }
13.   }
14.   iteration Update Issuezilla {
15.     action Report issues to Issuezilla {
16.       requires { Test results }
17.       provides { Issuezilla entry }
18.       tool { Web browser }
19.       agent { users, developers, Sun ONE Studio QA
            team, Sun ONE Studio developers }
20.     }
21.     action Update standing issue status {
22.       requires { Standing issue from Issuezilla, test
            results }
23.       provides { Updated Issuezilla issue repository }
24.       tool { Web browser }
25.       agent { users, developers, Sun ONE Studio QA
            team, Sun ONE Studio developers }
26.     }
27.   }
28.   action Post bug stats {
29.     requires { Test results }
30.     provides { Bug status report, test result report }
31.     tool { Web editor, JFreeChart }
32.     agent { Release manager }
33.   }
34. }
```

Figure 4. A PML description of the testing sequence of the NetBeans release process

As hinted earlier, because OSSD projects are open to contributions from afar, it also becomes possible to contribute explicit models of discovered processes back to the project under study so that project participants can openly review, independently validate, refine, adapt or otherwise improve their own software processes. Accordingly, we have contributed our process models and analyses of the NetBeans requirements and release process in the form of a public report advertised on the NetBeans.org Web site[6].

Process re-enactment allows us to simulate or prototype process enactments by navigationally traversing a semantic hypertext representation of the process[16, 24]. These re-enactment prototypes are automatically derived from a compilation of their corresponding PML process model[16]. One step in the process modeled for NetBeans appears in Figure 5.

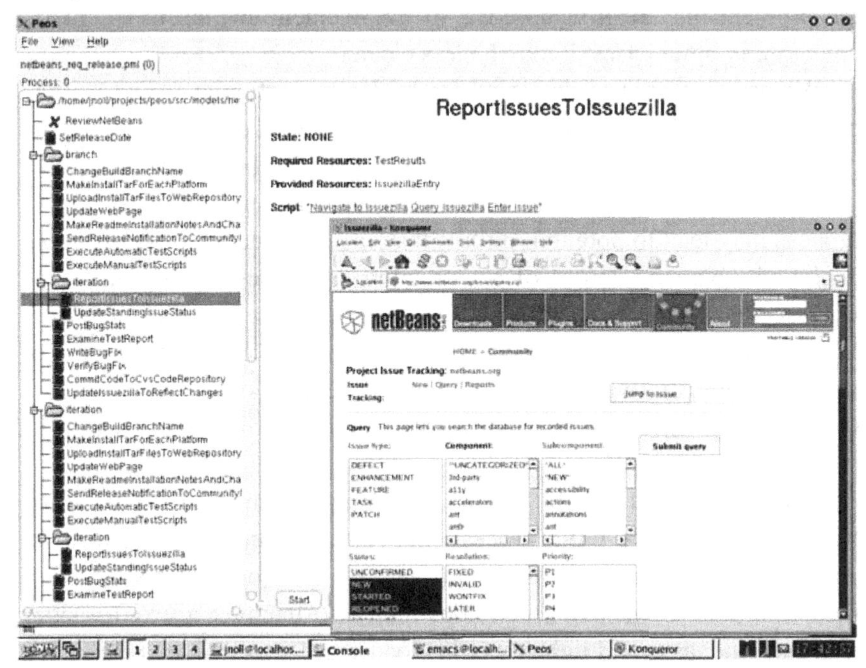

Figure 5. An action step in the re-enactment of the NetBeans requirements and release process

---

[6] See http://www.netbeans.org/community/articles/UCI_papers.html, as of October 2004

In exercising repeated process re-enactment walkthroughs, we have been able to detect process fragments that may be unduly lengthy, which may serve as good candidates for streamlining and process redesign[24]. Process re-enactment also allows us, as well as participants in the global NetBeans project, to better see the effects of their duplicated work. As an example, we have four agent types that test code. Users may carry out beta testing from a black box perspective, whereas developers, contributors, and SUN Microsystems QA experts may perform more in-depth white-box testing and analysis. In the case of developers and contributors, they will not merely submit a bug report or unsuccessful testing result to the IssueZilla issue tracking system,[7] but may also take responsibility for resolving it.

However, is it really necessary to have so many people doing such similar work? While, in this case, the benefits of having more eyes on the problem may justify the costs of involvement (which is voluntary, anyway), in other cases, it may be less clear.

We are also able to detect where cycles or particular activities may be problematic for participants, and thus where process redesign may be of practical value[24]. Process re-enactments can also be treated as process prototypes in order to interactively analyze whether or how altering a process may lead to potential pitfalls that can be discovered before being deployed. Over the course of constructing and executing our prototype we discovered some concrete reasons for why there are few volunteers for the release manager position. The role has an exceptional amount of tedious administrative tasks. However, as these tasks are critical to the success of the project it might be more effective to distribute these tasks to others.

Between scheduling the release, coordinating module stabilization, and carrying out the build process, the release manager has a hand in almost every part of the requirements and release process. This is a good indication that downstream activities may also uncover a way to better distribute the tasks and lighten her/his load.

The self-selective nature of OSSD project participation has many impacts on the development process in use. If any member does not want

---

[7] See http://www.netbeans.org/kb/articles/issuezilla.html, as of March 2004

to follow a given process, the enforcement of the process is contingent on the tolerance of her/his peers in the matter, which is rarely the case in corporate development processes. If the project proves intolerant of the alternative process, developers are free to simply not participate in the project's development efforts and perform an independent software release build.

7. Conclusion

Our desire is to obtain and model process execution data and event streams by monitoring the Web information spaces of open source software development projects. By examining changes to the information space and artifacts within it, we can observe, derive, or otherwise discover process activities. In turn, we reconstitute and abstract process instances into PML[16], which provides us with a formal description of an enactable, low-fidelity model of the process in question. Such a formal process model can be analyzed, simulated, redesigned, and refined for reuse and redistribution. But this progress still begs the question of how to more fully automate the discovery and modeling of processes found in large, global scale OSSD projects.

Our experience with process discovery in the NetBeans project, and its requirements and release process, suggests that a bottom-up strategy for process discovery, together with a top-down process meta-model acting as a reference model, can serve as a suitable framework for process discovery, modeling and re-enactment. As demonstrated in the testing activity example, action sequences are constructed much like a jigsaw puzzle. We compile pieces of evidence to find ways to fit them together in order to make claims about process enactment events, artifacts, or circumstances that may not be obvious from the individual pieces. We find that these pieces may be unearthed in ways that can be executed by software tools that are guided by human assistance[25].

The approach to discovery, modeling, and re-enactment described in this chapter relies on a variety of informal and formal process representations. We constructed use cases and rich hypermedia pictures as informal process descriptions, flow graphs as informal but semi-structured process representations which we transformed into a formal process representation language guided by a process meta-model and support tools. These informal representations together with a process meta-model then provide a basis for constructing formal process descriptions. Thus demonstration of a more automated process discovery, modeling, and re-enactment environment that integrates these capabilities and mechanisms into a more streamlined and more automated environment is the next step in this research. We anticipate that such an environment will yield additional venues for tool assistance in process data collection and analysis.

Finally, it is important to recognize that large OSSD projects are diverse in the form and practice of their software development processes. Our long-term goal in this research is to determine how to best support a more fully automated approach to process discovery, modeling and re-enactment. Our study provides a case study of a real-world process in a complex global OSSD project to demonstrate the feasibility of such an approach. Subsequently, questions remain as to which OSSD processes are most amenable to such an approach, and which are likely to be of high value to the host project or other similar projects. Furthermore, we need to establish whether all or only some OSSD projects are more/less amenable to such discovery and modeling given the richness/paucity of their project information space and diversity of artifacts. As government agencies, academic institutions and industrial firms all begin to consider or invest resources into the development of large OSS systems, then they will seek to find what the best OSSD processes are, or what OSSD practices to follow. Thus discovery and explicit modeling of OSSD processes in forms that can be shared, reviewed, modified, re-enacted, and redistributed appears to be an important topic for further investigation, and this study represents a step in this direction.

Acknowledgements

The research described in this report is supported by grants from the National Science Foundation #0083075, #0205679, #0205724, and #0350754. No endorsement implied. Mark Ackerman at the University of Michigan Ann Arbor; Les Gasser at the University of Illinois, Urbana-Champaign; John Noll at Santa Clara University; Margaret Elliott and others at the UCI Institute for Software Research are collaborators on the research described in this paper.

References

1. Mockus, A., Fielding, R. and Herbsleb, J. 2002. Two Case Studies in Open Source Software Development: Apache and Mozilla, *ACM Trans. Software Engineering and Methodology*, 11(3), 309-346.
2. *NetBeans Open Source Project*, 2003. http://www.netbeans.org
3. *Eclipse Web Site*, 2003. http://www.eclipse.org
4. Elliott, M. and Scacchi, W. 2004. Free Software Development: Cooperation and Conflict in A Virtual Organizational Culture, in S. Koch (ed.), *Free/Open Source Software Development*, Idea Publishing.
5. Scacchi, W. 2002. Understanding the Requirements for Developing Open Source Software Systems, *IEE Proceedings—Software*, 149(1), 25-39.
6. Scacchi, W. 2004. Free/Open Source Software Development Practices in the Game Community, *IEEE Software*, 21(1), 59-67, Jan-Feb. 2004.
7. Viller, S. and Sommerville, I. 2000. Ethnographically Informed Analysis for Software Engineers, *Intern. J. Human-Computer Interaction*, 53, 169-196.
8. Mi, P. and Scacchi, W. 1996. A Meta-Model for Formulating Knowledge-Based Models of Software Development, *Decision Support Systems*, 17(4), 313-330.
9. Cook, J. and Wolf, A.L. 1998. Discovering Models of Software Processes from Event-Based Data, *ACM Trans. Software Engineering and Methodology*, 7(3), 215-249.
10. Wolf, A.L. and Rosenblum, D.S. 1993. A Study in Software Process Data Capture and Analysis, *Proc. Second Intern. Conf. on the Software Process*, 115-124, IEEE Computer Society.
11. Garg, P.K. and Bhansali, S. 1992. Process programming by hindsight, *Proc. 14th Intern. Conf. Software Engineering*, 280-293.
12. Sandusky, R., Gasser, L., and Ripoche, G. 2004. Bug Report Networks: Varieties, Strategies, and Impacts in a F/OSS Development Community, *Proc. MSR'04 Workshop*, Edinburgh, Scotland, May 2004.
13. Lopez-Fernandez, L., Robles, G., Gonzalez-Barahona, J. 2004. Applying Social Network Analysis to the Information in CVS Repositories, *Proc. MSR'04 Workshop*, Edinburgh, Scotland, May 2004.
14. Cass, A.G., Lerner, B., McCall, E., Osterweil, L. and Wise, A. 2000. Little JIL/Juliette: A process definition language and interpreter, *Proc. 22nd Intern. Conf. Software Engineering*, 754-757, Limerick, Ireland, June.
15. Osterweil, L. 2003. Modeling Processes to Effectively Reason about their Properties, *Proc. ProSim'03 Workshop*, Portland, OR, May 2003.

16. Noll, J. and Scacchi, W. 2001. Specifying Process Oriented Hypertext for Organizational Computing, *Journal of Network and Computer Applications*, 24 39-61.
17. Cook, J., Votta, L. and Wolf, A.L. 1998. Cost-Effective Analysis of In-Place Software Processes, *IEEE Transactions on Software Engineering*, 24(8), 650-663.
18. Oza, M., Nistor, E., Hu, S. Jensen, C. and Scacchi, W. 2002. *A First Look at the Netbeans Requirements and Release Process.* http://www.ics.uci.edu/cjensen/papers/FirstLookNetBeans/
19. Jensen, C. and Scacchi, W. 2003. Applying a Reference Framework to Open Source Software Process Discovery, *Proc. 1st Workshop on Open Source in an Industrial Context*, OOPSLA-OSIC03, Anaheim, CA, October 2003.
20. Fowler, M. and Scott, K. 2000. *UML Distilled: A Brief Guide to the Standard Object Modeling Language*. Second Ed. Addison Wesley: Reading, MA.
21. Monk, A. and Howard, S. 1998. The Rich Picture: A Tool for Reasoning about Work Context. *Interactions*, 21-30, March-April 1998.
22. Noy, N.F., Sintek, M., Decker, S., Crubézy, M., Fergerson, R.W. and Musen, M.A. 2001. Creating Semantic Web Contents with Protégé-2000, *IEEE Intelligent Systems*, 16(2), 60-71.
23. Podorozhny, R.M., Perry, D.E. and Osterweil, L. 2003. Artifact-based Functional Comparison of Software Processes, *Proc. ProSim'03 Workshop*, Portland, OR, May 2003.
24. Scacchi, W. 2000. Understanding Software Process Redesign using Modeling, Analysis, and Simulation, *Software Process—Improvement and Practice*, 5(2/3), 183-195.
25. Jensen, C. and Scacchi, W. 2004. Data Mining for Software Process Discovery in Open Source Software Development Communities, submitted for publication.

Chapter 2

SOFTWARE PROCESS DYNAMICS: MODELING, SIMULATION AND IMPROVEMENT

Mercedes Ruiz[†], Isabel Ramos[‡], Miguel Toro[‡]

Department of Computer Languages and Systems
[†]Escuela Superior de Ingeniería
C/ Chile, 1. 11003 – Cádiz (Spain)
[‡]Escuela Técnica Superior de Ingeniería Informática.
Avda. Reina Mercedes, s/n. 41013 – Seville (Spain)
E-mail: mercedes.ruiz@uca.es
{isabel.ramos, miguel.toro}@lsi.us.es

The aim of this chapter is to introduce the reader to the dynamics of the software process, the ways to represent and formalize it, and how it can be integrated with other techniques to facilitate, among other things, process improvement. In order to achieve this goal, different approaches of software process modeling and simulation will be introduced, analyzing their pros and cons. Then, continuous modeling will be used as the modeling approach to build software process models that work in the qualitative and quantitative fields, assessing the decision-making process and the software process improvement arena. The integration of this approach with current process assessment models (such as CMM), static and algorithmic models (such as traditional models used in the estimation process) and the design of a metrics collection program which is triggered by the actual process of model building will also be described in the chapter.

1. Introduction

Worldwide, the demand for highly complex software has significantly increased in such a way that software has replaced hardware as having the principal responsibility for much of the functionality provided by current systems. The rapid pace at which this software is required, the problems related to cost and schedule overruns and customer perception of low product quality have changed the focus of attention towards the maturity of software development practices. Over the last few decades, the software industry has received significant help from CASE tools, new programming languages and approaches, and more advanced and complex machines.

However, it is widely accepted that the potential benefits of better technology cannot be translated into more successful projects if the processes are not well defined, established, and executed. Proper processes are essential for an organization to consistently deliver high quality products with high productivity.

Dynamic modeling and simulation have been intensively used as process improvement tools in the manufacturing area. Currently, interest in software process modeling and simulation as an approach for analyzing complex businesses and solving policy questions is increasing among researchers and practitioners. However, simulation is only effective if both the model and the data used to drive it accurately reflect the real world. As a consequence, it can be said that the construction of a dynamic model for the actual software process provides clear guidelines on what to collect.

Many frameworks are now available for software processes, the Capability Maturity Model (CMM)[1] and ISO 9001[2] being among the most influential and widely used. Although ISO 9001 is a standard, and has been interpreted for a software organization in ISO 9000-3[3], it has been written from the customer and external auditor's perspective. On the other hand, CMM is not a binary certification process, but a framework that categorizes the software process at five levels of maturity and provides roadmaps to evaluate the software process of an organization, as well as planning software process improvements. One of the common features that all these frameworks possess is that they strongly recommend the application of statistical control and measure guides to define, implement and evaluate the effects of different process improvements. Within these frameworks, the availability of data is considered of special importance for building the knowledge required to define and improve the software process.

The aim of this paper is to present a combination of traditional techniques with software process modeling and simulation to build a framework for supporting a qualitative and quantitative assessment for software process improvement and decision making. The purpose of this dynamic framework is to help organizations to achieve a higher software development process capability according to CMM. The dynamic models built within this framework provide the capability of gaining insight over the whole life cycle at different levels of abstraction.

The level of abstraction used in a particular organization will depend on its maturity level. For instance, in a level 1 organization the simulator can establish a baseline according to traditional estimation models from an initial estimate of the size of the project. With this baseline, the software manager can analyze the results obtained by simulating different process improvements and study the outcomes of an over- or underestimate of cost or schedule. During the simulation metric data is saved. This data conforms to the SEI core measures[4] recommendation and is mainly related to cost, schedule and quality.

The structure of the chapter is as follows. Section 2 describes in detail the software process modeling and simulation approach. It includes the benefits derived from this application, the formalisms used to build software process models and a process model building methodology. In section 3, a combination of hierarchical dynamic modeling and some traditional techniques of the software engineering is proposed. The conceptual ideas underlying this combination with the aim of building an integrated dynamic framework for software process improvement are presented. Sections 4, 5 and 6 describe the details concerning the structure of the framework, the modular architecture and some aspects of the implementation. An example of usage is presented in section 7. Finally, section 8 summarizes the chapter and describes the most recent applications of the software process dynamic modeling and simulation approach.

2. Software Process Simulation

Simulation can be applied in many critical areas in support of software engineering. It enables one to address issues before these issues become problems. Simulation is more than just a technique, as it forces one to think in global terms about system behavior and about the fact that systems are more than the sum of their components[5]. A simulation model is a computational model that represents an abstraction or a simplified representation of a complex dynamic system. The main benefit of simulation models is the possibility of experimenting with different management decisions. Thus, it becomes possible to analyze the effect of those decisions on systems where the cost or risks of experimentation make it unfeasible.

Another important factor is that simulation provides insights into complex process behavior that cannot be analyzed by means of stochastic models. Like many processes, software processes can contain multiple feedback loops, such as those associated with the correction of defects. Delays resulting from these defects may range from minutes to years. The resulting complexity makes it almost impossible for mental analysis to predict the consequences. According to Kellner, Madachy and Raffo[6], the most frequent sources of complexity in real software processes are:

- Uncertainty. Some real processes are characterized by a high degree of uncertainty. Simulation models make it possible to deal with this uncertainty as they can represent it flexibly by means of parameters and functions.
- Dynamic behavior. Some processes may have a time-dependent behavior. There is no doubt that the behavior of some software process variables varies as the time cycle progresses. With a simulation model it is possible to represent and formalize the structures and causal relationships that dictate the dynamic behavior of the system.
- Feedback. In some systems, the result of a decision made at a given time can affect their behavior. In software projects, for example, the decision to reduce the effort assigned to quality assurance activities has different effects on the progress of these projects.

Thus, the common objectives of simulation models are to supply mechanisms to experiment, predict, learn and answer questions, such as "What if ...?"

A software process simulation model can be focused on certain aspects of the software process or the organization. It is important to bear in mind that a simulation model constitutes an abstraction of the real system, and so it represents only the parts of the system that were intended to be modeled. Furthermore, currently available modeling tools, such as ithink[7], POWER-SIM[8], and Vensim[9], help to represent the software development process as a system of differential equations. This is a remarkable characteristic as it makes it possible to formalize and develop a scientific basis for software process modeling and improvement.

During the last decade, software process simulation has been used to address a wide variety of management problems. Some of these problems are related to strategic management, technology adoption, understanding, training and learning, and risk management, among others. Noticeable applications of this approach to software process modeling can be found in Kellner, Madachy and Raffo[6], Prosim 2004[10] and Prosim 2005[11].

2.1. *Software process modeling for simulation*

There are different approaches for building simulation models of the software process. In practice, the modeling approach inevitably has some influence on what it should be modeling. Hence, there is no preferred approach for modeling the software process in every situation, but the best approach is always the one that is considered to be the most suitable for a particular case.

There are two broad types of simulation modeling: continuous simulation and discrete-event simulation. The distinction is based on whether the state can change continuously or at discrete points in time. However, even though events are discrete, time and state domains may be continuous. There are three main paradigms that can be used for discrete-event simulation modeling: event-scheduling, activity-scanning and process-interaction. Although state-transition diagrams (e.g., finite-state automata or Markov chains) can be used for software process simulation modeling, they are less common because the state spaces involved are typically very large. Examples of discrete-event simulation applied to model and simulate the software process can be found in Raffo[12], Kellner[13] and Hansen[14].

A continuous simulation model represents the interactions between key process factors, as a set of differential equations, where time is increased step by step. Frequently, the metaphor of a system of interconnected tanks filled with fluid is used to exemplify the ideas underlying this kind of modeling approach.

On the other hand, discrete modeling is based on the metaphor of a queuing network where time advances when a discrete event occurs. When this happens, an associated action takes place, which, mostly, implies placing a new event in the queue. Time is always advanced to the next event, so it can be difficult to integrate continually changing variables.

Since the purpose of this study is to model and visualize process mechanisms, continuous modeling has been used. This technique also allows systems thinking and it is considered to be better than the discrete-event model at showing qualitative relationships[15]. Examples of continuous simulation applied to model and simulate the software process can be found in Abdel-Hamid[16], Pfhal and Lebsant[17], Burke[18], and Wernick and Hall[19].

2.2. *Continuous modeling and simulation of the software process*

System dynamics is a methodology for studying and analyzing complex feedback systems such as software organizations. Feedback is the key differentiating factor of dynamic systems. It refers to the situation in which A affects B and B affects A, through a chain of causes and effects. It is not possible to study the link between A and B and, independently, the link between B and A to predict the behavior of the system. There are a significant number of software process features that follow this feedback pattern. For instance, known patterns, such as Brook's Law[20] ("Adding manpower to a late project makes it later") or Parkinson's Law[21] ("Work expands to fit the time available"), can be described by continuous modeling.

System dynamics links structure (feedback loops) to behavior over time and helps to explain *why what* is happening *is* happening. The field was initially developed from the work of Jay W. Forrester[22].

To better understand and represent the system structures that cause the patterns of behavior observed in the software process, two kinds of diagrams are used: causal-loop diagrams and stock-and-flow diagrams.

2.2.1. *Causal-Loop Diagrams*

Causal-loop diagrams present relationships that are difficult to describe verbally because natural language presents interrelations in linear cause-and-effect chains, whereas a diagram shows that there are *circular* chains of cause-and-effect in the actual system[23]. Figure 1 shows an example of a causal-loop diagram for a very simplified model of software process dynamics. In this diagram, the short descriptive phrases represent the elements that make up the system described, and the arrows represent the causal influences between these elements. This diagram includes elements and arrows or links that help to connect these elements, but also includes a sign (either + or -) on each link. These signs have the following meaning[23]:

- A causal link from one element A to another element B is *positive* if either (a) A adds to B or (b) a change in A produces a change in B in the *same* direction.
- A causal link from one element A to another element B is *negative* if either (a) A subtracts from B or (b) a change in A produces a change in B in the *opposite* direction.

In addition to the signs of each link, a complete loop is also given a sign. The sign of a particular loop is determined by counting the number of minus signs on all the links that make up the loop. Specifically,

- A feedback loop is called *positive*, indicated by (+), if it contains an even number of negative causal links.
- A feedback is called *negative*, indicated by (-), if it contains an odd number of negative causal links.

Thus, the sign of a loop is the algebraic product of the signs of its links. The diagram shown in Figure 1 is composed of four feedback loops: two positive and two negative. A brief description of the pattern modeled follows.

First feedback loop. Estimations of cost and time for the project can be derived from the initial estimations. With these estimations the required manpower is acquired by performing hiring activities. As the project runs, information about the real progress is obtained. Comparisons of the values obtained with those originally estimated may lead to a change in some of the estimations and, possibly, a modification of the hiring policy.

Second feedback loop. This loop illustrates the effects caused by the schedule pressure on the quality of the software product. If the perceived completion time is greater than the planned time to complete, the project has schedule pressure. To combat this, the project manager may decide either to hire more personnel or have overtime worked. However, permanent overtime may further exhaust personnel, contributing to an increase in the number of errors in the project. This rise in the number of committed errors requires a bigger effort in terms of error detection and rework activities, which holds back progress.

Third feedback loop. The growth in the level of human resources appears to contribute to a growth of productivity. However, it is also important to note that the productivity of the new personnel is significantly less than that of the expert personnel. Hence, some effort of the expert personnel is commonly invested in the training of the newly-hired personnel. These training activities, together with the communication overheads derived from the Book's Law, contribute to a decrease in the net productivity of the working team.

Fourth feedback loop. This loop illustrates the effect of creative pressure. When the personnel know that the project is behind schedule, they tend to be more efficient. This is normally reflected in a reduction of idle time.

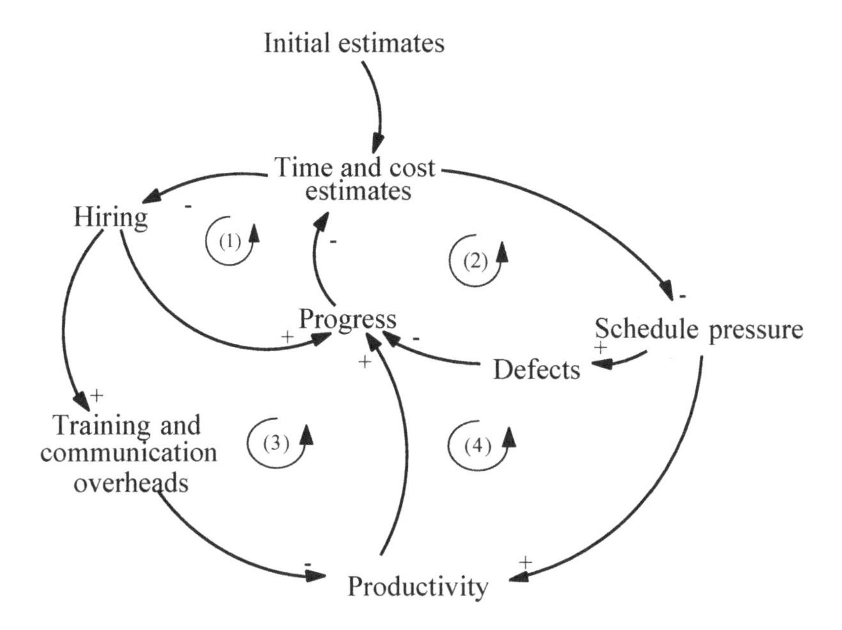

Fig. 1. Simple causal-loop diagram of the software process dynamics.

2.2.2. *Stock-and-Flow Diagrams*

Figure 2 illustrates the main components of stock-and-flow diagrams. This notation consists of three different types of elements: stock, flows and information. These three elements provide a general way of graphically representing *any* process. Furthermore, this graphical notation can be used as a basis for developing a quantitative model that can be used to study the characteristics of the process. As with a causal-loop diagram, the stock-and-flow diagram shows relationships among *variables* that have the potential to change over time. To understand and build stock-and-flow diagrams, it is necessary to understand the difference between stocks and flows. Distinguishing between stocks and flows is sometimes difficult. As a starting point, stocks can be thought of as physical entities that can accumulate and move around. The term *stock* also has an identical meaning to the term *state variable* from the systems engineering analysis. The term *flow* refers to the movement of something from one stock to another.

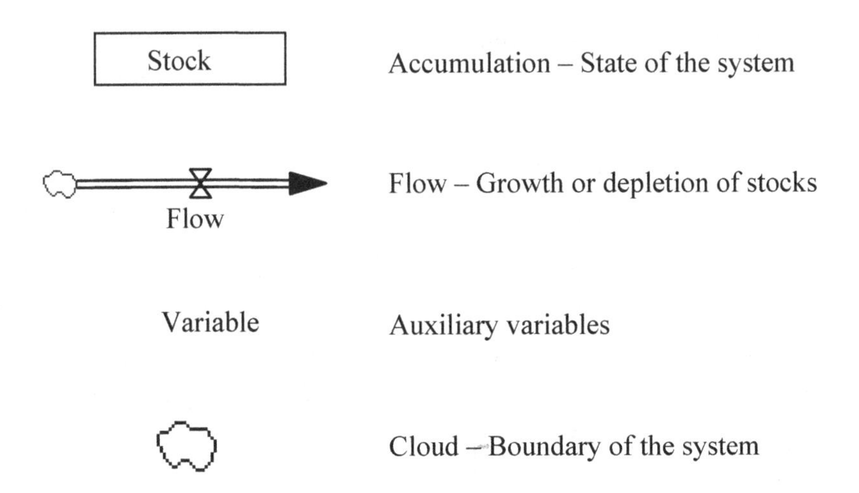

Fig. 2. Main elements of stock-and-flow diagrams.

Figure 3 shows a stock-and-flow diagram for the first feedback loop of the causal diagram shown in Figure 1. The variables are *Pending tasks*, *Accomplished tasks*, *Personnel*, *hiring rate* and *development rate*. The first three are stock or level variables, whereas the last two are flow variables. The number of tasks to be developed is determined from an initial estimate of the size of the project. These pending tasks become accomplished tasks depending on the development rate that is determined by the productivity of the personnel allocated to the development of the tasks under simulation.

The stock-and-flow diagram has a precise mathematical meaning. Stocks accumulate (integrate) their inflows less their outflows. The rate of change of a stock is the total inflow minus the total outflow. Thus a stock and flow map corresponds to a system of integral or differential equations that formalize the model. Mathematical description of a system requires only the stocks and their rates of change. However, it is often helpful to define intermediate or auxiliary variables. Auxiliaries consist of functions of stocks and constants. The set of equations must then be solved applying mechanisms for solving differential equations or by simulation. Simulation packages are often used to solve these sets of equations, since it soon becomes unfeasible to solve such equations by hand as the number of stocks and flows or the complexity of the equations increases.

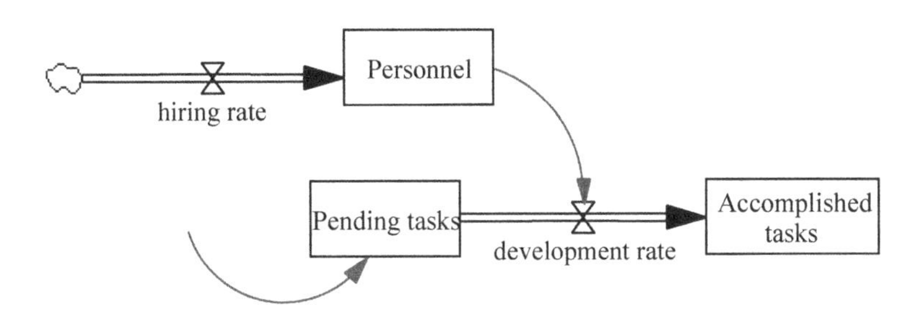

Fig. 3. Simple stock-and-flow diagram.

The equations derived from the stock-and-flow diagram follow:

$$\text{Pending tasks(t)} = \text{INITIAL SIZE ESTIMATES} - \int_0^t \text{development rate(t) dt} \quad (1)$$

$$\text{Accomplished tasks(t)} = \int_0^t \text{development rate(t) dt} \quad (2)$$

$$\text{Personnel(t)} = \int_0^t \text{hiring rate(t) dt} \quad (3)$$

$$\text{development rate(t)} = \begin{cases} \text{Personnel(t)} * \text{Productivity(t), if Accomplished tasks(t)} < \text{INITIAL SIZE ESTIMATES} \\ \text{0, otherwise} \end{cases} \quad (4)$$

$$\text{hiring rate(t)} = \left(\text{required personnel(t) - Personnel(t)}\right)/\text{HIRING DELAY} \quad (5)$$

Figure 4 shows the time evolution of the main variables of this illustrative model after solving the equations by simulation.

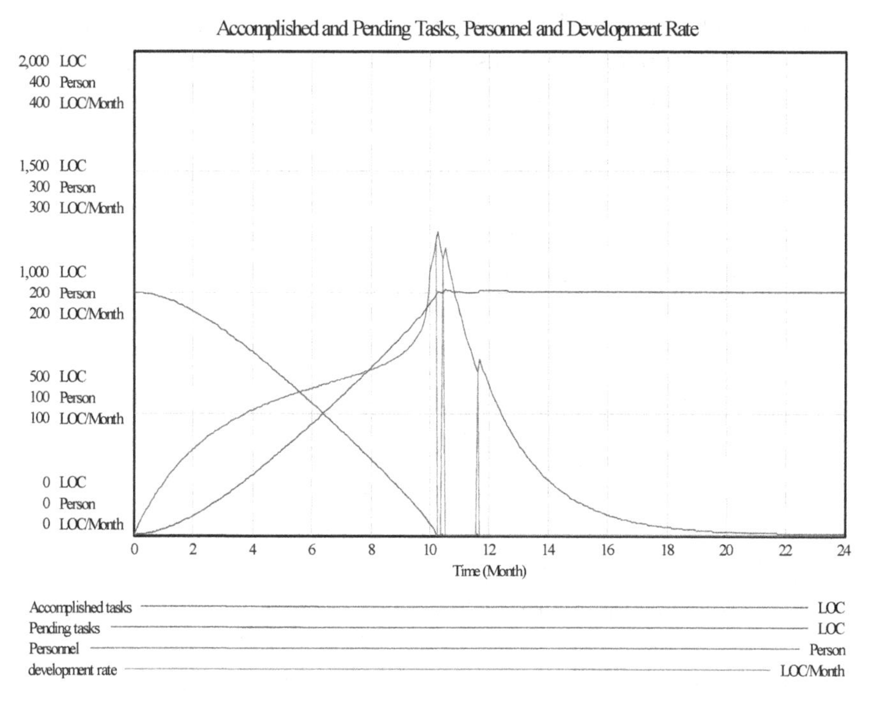

Fig. 4. Time evolution of the main variables of the stock-and-flow diagram.

Nevertheless, as Sweeny and Sterman[24] stated, building a model is not about spending considerable time on the basics of stocks and flows, time delays, and feedback, but developing intuition rather than mathematics.

2.3. *Process model building methodology*

According to Martinez and Richardson[25], the system dynamics model building process involves seven key activities, as shown in Figure 5. The most important ones are: (1) problem identification and definition, (2) system conceptualization, (3) model formulation, (4) model testing and evaluation, and (5) understandings of the model.

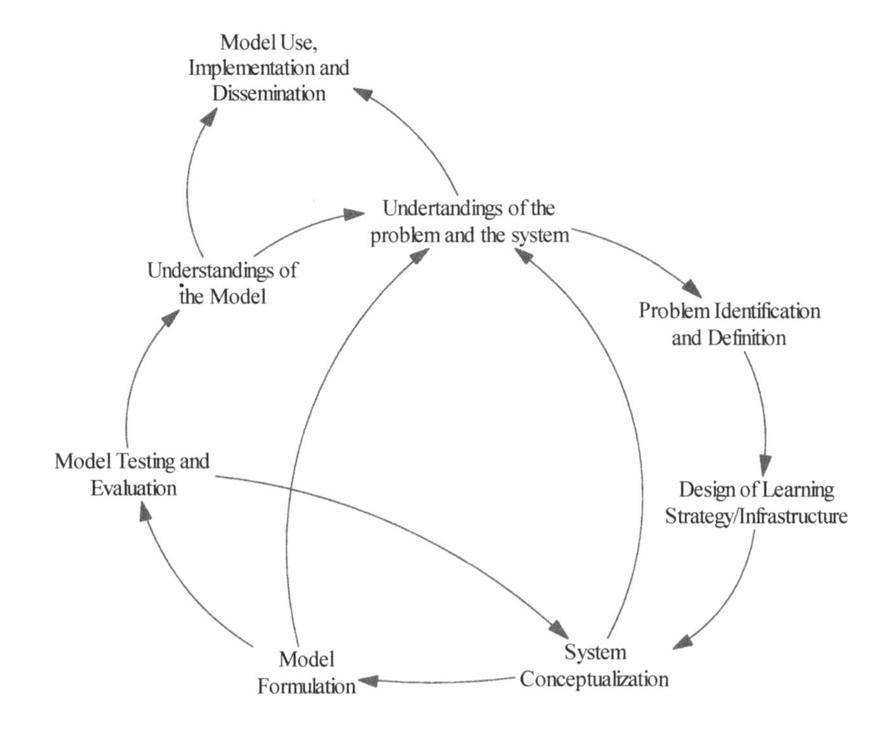

Fig. 5. Steps of process model building methodology.

In *problem identification and definition*, there is a group of practitioners who consistently prefer to model the case at hand, as opposed to another group who thinks that the best way is to model the class to which the system belongs.

In *system conceptualization*, the best practice is considered to be to start with major stock variables. Practitioners can choose to iterate using a causal-loop diagram approach or a stock-and-flow approach to conceptualization.

In *model formulation*, there are two major approaches. The first relates to the issue of starting small and continuously simulating, preferably always having a running model. The second refers to formulating in big chunks and is not concerned about continuously having running prototypes.

Model testing and evaluation consists of three main activities that determine the correctness of the model[26]. These activities are divided into two categories: activities focused on verifying the model structure and activities that verify the model behavior. Table 1[27] summarizes these activities.

Finally, *understandings of the model* is centered on the knowledge that can be gained from use of the model.

3. Dynamic Integrated Framework for Software Process Improvement: Conceptual Approach

Using simulation for process improvement in conjunction with CMM is not a new idea. As a matter of fact, Christie[5] suggests that CMM is an excellent incremental framework to gain experience through process simulation. Nevertheless, there are no dynamic frameworks capable of helping to achieve higher process maturity. One of the main features of the Dynamic Integrated Framework for Software Process Improvement (DIFSPI) is that this help is provided throughout the development of the whole dynamic framework and not only by using the associated final tool. The reason for this is that the benefits that can be gained from the utilization of dynamic models within an organization are directly related to the knowledge and the empirical information that the organization has about its processes. Figure 6 illustrates this idea. It shows the existing causal relationships between the maturity level of the organization, the utilization of dynamic models and the benefits obtained.

The positive feedback loop comes to illustrate the causal relationship that reinforces the collection of metrics within the organization. The metrics collected will be used to calibrate and initialize the dynamic models.

Lower maturity organizations are characterized by the absence of metrics programs and historical databases. In this case, it is necessary to begin by identifying the general processes and what information has to be collected about them. The questions of what to collect, how often and how accurately have to be answered at this time. The design process of dynamic models helps to find an answer to these questions.

Table 1. Main model testing and evaluation activities[27].

Verification	Structure	Dimensional consistency
		Behavior with extreme values
		Problem adequacy
	Behavior	Parameter sensitivity
		Structure sensitivity
Validation	Structure	Reality check
		Parameter correctness
	Behavior	Scenario replication
		Extreme condition simulations
		Non-conventional input simulations
		Statistical tests
Evaluation	Structure	Size
		Complexity
		Granularity
	Behavior	Intuitive behavior generation
		Knowledge generation

When developing a dynamic model, one needs to know: a) what it is intended to model, b) the scope of the model, and c) what behaviors need to be analyzed.

Once the model has been developed, it needs to be initialized with a set of initial conditions in order to execute the runs and obtain the simulated behaviors. These initial conditions customize the model to the project and to the organization to be simulated and they are effectively implemented by a set of initial parameters.

It is precisely these parameters that govern the evolution of the model runs that answer the above question of what data to collect: the data required to initialize and validate the model will be the main components of the metrics collection program. Once the components of the metrics collection program have been derived, it can be implemented within the organization. This process will lead to the formation of a historical database. The data gathered can then be used to simulate and empirically validate the dynamic model. When the dynamic model has been validated, the results of its runs can be used to generate a database. This database can be used to perform process improvement analyses. An increase in the complexity of the actions for analysis will lead directly to an increase in the complexity of the dynamic model required and, therefore, to a new metrics collection program for the new simulation modules.

The bottom half of Figure 6 illustrates the effects derived from the utilization of dynamic models in the context of process improvement. Using dynamic models that have been designed and calibrated according to an organization's data has three important benefits. Firstly, the data from the simulation runs can be used to predict the evolution of the project. The graphical representations of these data show the evolution of the project from a set of initial conditions that have been established by the initialization parameters. By analyzing these graphs, organizations with a low level of maturity can obtain useful qualitative knowledge about the evolution of the project. As the maturity level of the organization increases, the knowledge about its processes is also higher and the simulation runs can be used as real quantitative estimates. These estimates help to predict the future evolution of the project with an accuracy that is closely related to the uncertainty of the initial parameters. Secondly, it becomes possible to define and experiment with different process improvements by analyzing the different simulation runs. This capability helps in the decision-making process, as only the improvements that yielded the best results will be implemented. Moreover, it is noteworthy that these experiments are performed at no cost or risk to the organization, as they use the simulation of scenarios. Thirdly, the simulation model can also be used to predict the cost of the project; this cost can be referred to the overall cost, or to a hierarchical decomposition of the total cost, like, for instance, the cost of quality or rework activities. These three benefits are the main factors that lead to the achievement of a higher maturity level within an organization according to CMM.

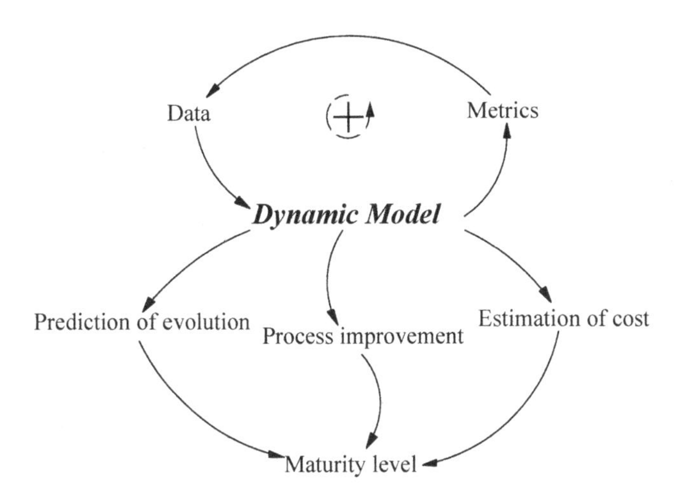

Fig. 6. Causal relationships concerning the utilization of dynamic models.

4. Framework Structure and Module Architecture

Project management is composed of activities that are closely interrelated in the sense that any action taken in one particular area will possibly affect other areas. For instance, a time delay will always affect the cost of the project, but it may or may not affect the morale of the development team or the quality of the product. The interactions among the different areas of project management are so strong that sometimes the throughput of one of them can only be achieved by reducing the throughput of another. A clear example of this behavior can be found in the common practice of reducing the quality, or the number of requirements to be implemented in a version of the product with the aim of meeting the cost estimates or time deadlines.

Dynamic models are an aid for understanding the integrated nature of project management, as they describe it by means of different processes, structures, and key interrelationships.

In the framework proposed here, project management is considered as a set of dynamic interrelated processes. Projects are composed of processes. Each process is composed of a series of activities designed to achieve an objective[1]. From a general point of view, it could be said that projects are composed of processes that fall into one of the following categories:

- Management process. This category includes all processes related to the description, organization, and control of the project.
- Engineering process. All processes related to the software product specification and development activities fall into this category.

Engineering processes begin to be executed from an initial plan performed by the project management processes. Using the information gathered about the progress of this second group of processes, project management processes determine the modifications that need to be made to the plan in order to achieve the project objectives. The proposed DIFSPI follows this same classification and is structured to account for project management and engineering processes. At both levels, the utilization of dynamic models to simulate real processes and to define and develop a historical database will be the main feature.

4.1. *Engineering processes in the DIFSPI*

At this level the dynamic models simulate the life cycle of the software product. In low maturity organizations, the amount of information required to begin running simulations is relatively small and mainly focused on the initial estimations, that is, the estimated size of the project and the initial size of the working team. The best dynamic model is simulated depending on the paradigm followed to develop the software product and the maturity level of the organization. The main paradigms that can be currently simulated within the framework are the traditional waterfall and COTS paradigms. Depending on the chosen paradigm, different dynamic modules will be joined in order to create a final and fully operational dynamic model. Once the simulation has been run, it provides data that are saved in a database. This initial data contains the results of the simulation together with a set of initial estimations resulting from the computation of the static models. These initial estimations establish the baseline for the project, and the simulated data obtained represent the dynamic evolution of the project variables throughout the whole life cycle. Apart from storing the initial baseline and the simulated data, the database contains a third component. This third component contains the results of applying some other techniques during the simulation of the project, which are oriented towards gaining insight into the process under simulation. These techniques, which have been integrated with the dynamic modules, are described in section 5.

As mentioned before, the process of modeling the software process requires a good knowledge of the software process itself, and triggers a metrics collection program that can then be used to initialize the parameters of the model and increment the level of visibility the model has of the process. All that has been simulated so far must be put into practice.

After determining the initial estimates and running the simulations to establish the initial baseline, it is possible to run different scenarios in order to find out what effects different initial values have on the project estimates. This reflects, of course, the level of uncertainty that low maturity organizations have at the initial stages of a project. When the real project begins, the metrics collection program may be applied to gather real information about the process. This real data is also saved in the database, enabling the development of a historical database. As this data becomes available, it is possible to perform analysis and calibrate the functions and parameters of the dynamic modules so that their accuracy can be improved. Improving the accuracy of the dynamic modules may require an improvement in the knowledge we have of the software process and, this way, the loop is closed.

The dynamic models of this level of DIFSPI should follow the levels of visibility and knowledge of the engineering processes that organizations have at each maturity level. Obviously the dynamic model used in level 1 organizations will not be as complex as the models capable of simulating the engineering processes of, for instance, level 4 organizations.

4.2. *Management processes in the DIFSPI*

The control modules model and simulate all the activities that determine the progress of the project, and make the corrective decisions that are required to meet the project objectives. These modules are highly important in the design of the process improvements.

Within the framework, management processes are divided into two main categories:

- Planning. It groups the processes devoted to the design of the initial plan and the required modifications when the progress reports indicate the appearance of problems. The models of this group integrate traditional together with dynamic estimation and planning techniques.
- Control. This group includes all the models designed for monitoring and tracking activities. These models will also have the responsibility of determining the corrective actions to the project plan. Therefore, the simulation of process improvements will be of enormous importance.

Figure 7 shows the utilization of DIFSPI at this level. As mentioned earlier, the initial baseline for the project is established using the static models built within the framework. The dynamic modules that model the planning activities performed in the organization not only have differential equations to model these activities, but also the equations of the traditional static estimation models. To gain useful information from these static models, the very same knowledge about the software process is needed at this point as is required to use these models.

4.3. *Module architecture*

The approach followed to construct the dynamic models is based on two fundamental principles:

The principle of *extensibility* of dynamic models. According to this principle, different dynamic modules are joined to an initial and basic dynamic model. This initial model models the fundamental behavior of a software project. Each one of the dynamic modules models each one of the key process areas that conforms the step to the next level of maturity. These modules can be either "enabled" or "disabled" according to the objectives of the project manager or the members of the Software Engineering Improvement Group (SEIG).

The principle of *aggregation/decomposition* of tasks according to the level of abstraction required for the model. Two levels of aggregation/decomposition are used:

- Horizontal aggregation/decomposition according to which different sequential tasks are aggregated into a unique task with a unique schedule.
- Vertical aggregation/decomposition according to which different and individual, but interrelated and parallel tasks are considered as a unique task with a unique schedule too.

The definition of the right level of aggregation and/or decomposition for the tasks mainly affects the modeling of the engineering activities and principally depends on the maturity level of the process to be simulated.

To define the initial dynamic model, the common feedback loops among the software projects must be taken into account. The objective of this approach is to achieve a generic model avoiding the modeling of specific behaviors of concrete organizations, which could limit the flexibility of DIFSPI. Data from historical databases described in the available literature can be used to initialize the functions and parameters of the initial model[28]. Figure 7 shows the main structure of the initial model. Four dynamic modules are joined together to develop an operational model that provides the set of final differential equations to generically simulate the software process in low maturity organizations.

By replicating some of the equations of the initial model it is possible to model the progress to higher maturity levels. The initial model can be used to simulate software projects developed in organizations progressing to level 2.

Generally speaking, the software product development process can be considered as follows. The number of tasks to be developed is determined from an initial estimate of the size of the project. These pending tasks become accomplished tasks according to the development rate. During this process, errors can be committed. Thus, in accordance with the desired quality objective for the project, the quality rate and the rework rate are determined. These two rates govern the number of tasks that are revised.

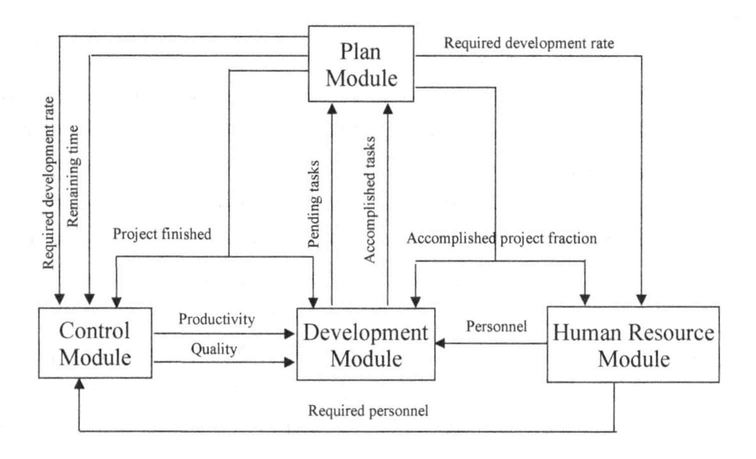

Fig. 7. Submodules architecture of the initial model.

To model the progress to level 3, the model will make use of a horizontal decomposition, creating as many substructures as phases or activities are present in the task breakdown structure of the project (analysis, design, code and test, in the waterfall paradigm). According to this approach, each time a complete model or some part of it is replicated, it will be necessary to define the new fixing mechanisms (dynamic modules) for the new structures. These mechanisms effectively implement the above-mentioned principle of aggregation/decomposition.

The replication of structures also provides the possibility of replicating the modules related to the project management processes. This replication is especially useful for high maturity level organizations, which will be able to establish process improvement practices for each particular activity of the life cycle.

Having described the approach to the elaboration of the dynamic models, this section gives a description of the hierarchical structure of the framework presented in this paper.

Figure 8 illustrates this hierarchy. The dynamic model for level 1 organizations progressing to level 2 is composed of four main dynamic modules, each of them devoted to modeling and simulating each of the four main subsystems of the software process: planning, human resource management, control, and development activities. These four subsystems form an initial dynamic model. This initial model is intended to be used in level 1 organizations progressing to level 2.

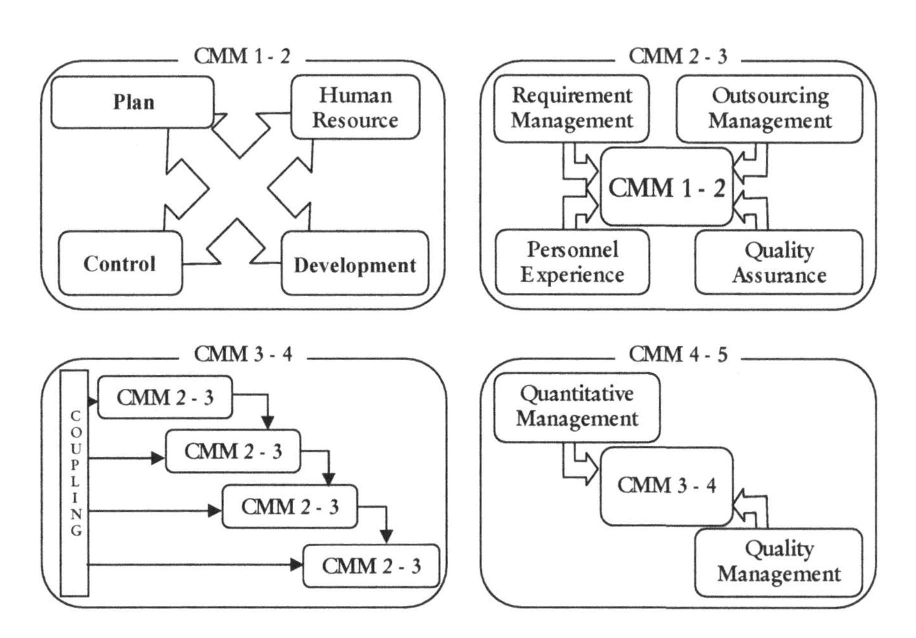

Fig. 8. Hierarchical structure of the dynamic integrated framework.

To get a dynamic model to model and simulate the software process of level 2 organizations, new dynamic modules are added to the initial model.

Outsourcing management. With this module, it is possible to analyze the influence of outsourcing over the life cycle of the project.

Personnel experience. Although this is not a key process area of CMM level 2, the human resource management module of the initial model has been enhanced so that it can reflect the influence of the experience factor on the progress and the cost of the project.

Quality assurance. The necessary structures to model and analyze the cost and state of the quality assurance activities are implemented in this module.

Requirement management. This module helps to determine the impact of requirements variability on software development projects.

The next step towards the following level of maturity does imply an important structural change. This change is determined by the special emphasis on the engineering activities that the CMM suggests as of level 3. While the CMM recommends the development of good planning and management practices in the initial levels of maturity, the engineering process acquires key importance at level 3. The principle of model replication is used to reflect this idea. Thus, to model level 3 organizations progressing to level 4, the model developed for the previous level is replicated as many times as the number of generic phases there are in the work breakdown structure of the project. For the purpose of this study, the four main characteristic phases of a traditional life cycle were considered (analysis, design, code and test). To simulate each phase, a complete dynamic model is used. Each of these dynamic models can be used, separately, to simulate the whole project in organizations with the previous level of maturity. To get all these models working together to simulate a higher maturity organization, coupling structures need to be defined. These coupling structures must allow inter-module communication as well as serving as support structures for the sharing of information.

The last model of the hierarchy is made from the model developed for the previous level, plus the modules required to model and simulate the new key process areas. In this case, the new modules are focused on the specific aspects of the key processes of quantitative management and software quality management.

5. Integrated Techniques

As mentioned before, our aim was to develop a working environment where the simulation of different scenarios can be used to generate the simulated database where managers can experiment with different process improvements and activities focused on the implementation of metrics programs and value analysis. The following techniques and methods are currently successfully implemented in DIFSPI:

Traditional estimation techniques. Traditional algorithmic estimation models have been implemented within this framework with the aim of providing an initial baseline for software projects carried out in low maturity level organizations [29, 30].

SEI Core Measures. Recent studies and experiences highlight the benefits of the application of these four core measures to the software life cycle. The main aspects of the product and process (quality, time, size and cost) are monitored and tracked to facilitate project success and higher maturity achievement. Within this framework these four measures constitute the basics for both the dynamic models and the graphical representation of process performance[4].

Metrel Rules. Given the dynamical nature of the proposed DIFSPI, we consider it could be useful to integrate a taxonomy of software metrics derived from the needs of users, developers, and management. Of all the potential advantages of using this system of metrics, we would like to point out the dynamic performance of these metrics, that is, how their accuracy, precision, and utility changes throughout a project, the life of a product or the strategic plan of an organization. In DIFSPI Metrel rules have been used as an efficient method for depicting on one graph the information needed for management, staff, and customers to view or predict process performance results. We consider that Metrel rules are particularly important in the field of software process modeling as their application provides a formal procedure for the expansion and transformation of models. By employing simple mechanisms like derivatives or integration (over time, phases or even projects), a mathematical model for one level can be transformed into another for another level, providing a simple but powerful extension for the analysis processes[31].

CoSQ. The basis for the Cost of Software Quality (CoSQ) is the accounting of two kinds of costs: costs that are due to a lack of quality and costs that are due to the achievement of quality. We think that CoSQ can help not only to justify quality initiatives, but also have a number of other benefits. Of these benefits, we would like to point out that CoSQ provides the basics for measuring and comparing the cost effectiveness of the quality improvements undertaken by an organization[32, 33].

Earned value analysis. Earned value analysis has been chosen as the method for performance measurement as it integrates scope, cost, and schedule measures to help managers assess process performance. The three main values and the derived efficiency indexes are used in combination to provide measures of whether or not work is being accomplished as planned. Furthermore, the earned value analysis is used to evaluate the performance of different software process improvements within DIFSPI[34] .

Statistical process control. Current software process models (CMM, SPICE, etc.) strongly recommend the application of statistical control. In the framework, Statistical Process Control (SPC) is used to obtain run charts and control charts with the aim of helping software managers to find an answer to questions such as "How do I know if my software development process is under control?" SPC is also used to test the capability of the process. For this purpose, SPC and earned value techniques can be merged as Lipke and Jennin[35] suggest.

Data mining. Data mining processes can be used to get useful information from the volume of data generated by model simulation. Genetic algorithms are fed with the databases resulting from simulations, and then executed to obtain management rules to guide the process of maturity improvement[36]. Machine learning algorithms based on decision trees such as C4.5[37], decision lists such as COGITO[38], and association rules[39] have been used in combination with other algorithms that transform the simulation outputs into a labeled database. In this labeled database, each record stores information about one simulated scenario (parameters and outputs) and a label that helps to classify the success of the simulated project in terms of time, cost, and quality. After running the machine learning algorithms, a set of management rules is obtained. These rules can be expressed graphically or using natural language. The information they offer is what the best range for the parameters that the algorithm has determined to be the most influential on the success of the project should be to meet the objectives of the project. These objectives, regarding the three key factors of time, cost, and quality together with the labeled databases, constitute the input of the algorithm.

6. Implementation of the Framework

The conceptual ideas presented above were firstly implemented using VemSim® which was used to develop and analyze the different dynamic models. However, there are some drawbacks to using this tool. This simulation environment provides a crude way of modularization, there is no easy way to both overlay objects for abstraction and generate a generic sub-model so that it can be instantiated multiple times without duplicating effort, and hence there is no scoping mechanism, all the elements are global to each other. Like traditional programming languages, a mechanism to allow data encapsulation and modularity is essential for handling complexity in large and complex models. Therefore, the complete framework has been re-engineered using UML and Java™ technology. The purpose of this process was to develop a library of classes, each of which represents a simple dynamic module. When using this tool, a suitable dynamic model is built from the required objects. This way, the abstraction aspect and standardization of the interface of these defined modules may be taken to the point that project managers could transparently "plug-in" the modules regarding the software process improvement they would like to analyze. This approach involves putting special effort into the interfacing mechanism of these different modules when they are plugged together.

7. Example of Usage

This section contains an example of how the use of this framework can help organizations in the field of software process improvement. More precisely, the following example studies one of the key process areas of CMM level 2: influence of the outsourcing activities on software projects. Table 2 shows the initial data for the project.

Table 2. Initial estimates for the project.

Size	20 KLDC
Number of newly hired engineers	3 engineers
Number of expert engineers	5 engineers
Estimated time	35 months
Number of outsourced tasks	150 tasks
Loss of effort due to outsourcing (%)	15%
Project reduction (%)	5%

Given this initial situation, two different scenarios are simulated. Both of them have the same initial data except for outsourcing activities: one of the projects does not have any outsourcing activities, while the other one does and is driven by the data shown in Table 2. The results obtained from the simulation runs are shown in the following subsections.

7.1.1. *Accomplished Tasks*

Figure 9 shows the evolution of tasks accomplishment in the project. First of all, it can be observed that the development rates in both projects are of a similar shape. Secondly, the project with outsourcing ends before the project without outsourcing. This may be due to the fact that the organization with outsourcing is carrying out a project that is smaller in size than the project of the organization that is not outsourcing. The vertical dotted line shows when the project with outsourcing is completed.

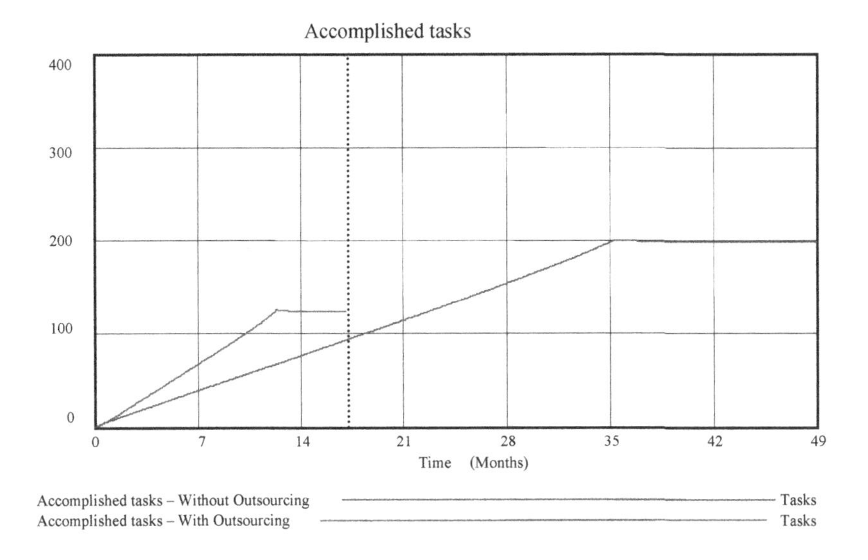

Fig. 9. Evolution of the variable *Accomplished tasks.*

7.1.2. *Effort*

Figure 10 shows the evolution of the daily effort consumed in the project activities. Notice that the effort values, and therefore cost, for the project with outsourcing are greater than the values for the project without outsourcing.

These higher costs are justified by the effort that needs to be allocated to some activities that are not present in the second project. When a project has outsourcing, some effort has to be allocated to mainly formal communication activities with the members of the outsourced team. This effort allocation leads to the growth of the final costs, a feature that has been illustrated by the simulation outputs.

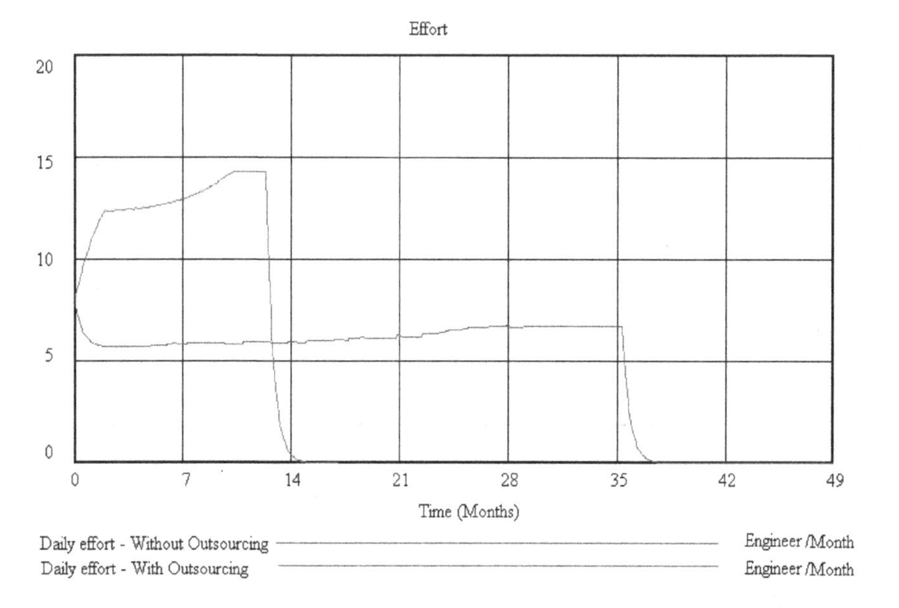

Fig. 10. Evolution of the variable *Daily effort.*

7.1.3. *Quality*

Finally, Figures 11 and 12 illustrate the aspects concerning the quality of the product under development. The initial quality objective for both projects is set as the number of tasks that need to be demonstrated, tested and corrected. This percentage is 90% for both projects. Figure 11 shows that this percentage is maintained for most of the duration of the lifecycle. However, when the final phase of each project is close, the percentage of tested tasks diminishes considerably. Nevertheless, the project with outsourcing achieves a higher level of final quality. The explanation for this result can be found in Figure 12. Figure 12 shows the evolution of the error detection rate. It can be observed that the project with outsourcing has a much higher error detection rate than the project without outsourcing. This behavior may be due to the fact that in the project with outsourcing, part of the quality assurance activities is performed by the outsourced team. Hence, the volume of tasks that need to be tested, demonstrated and corrected within the organization is significantly lower, and this makes it possible to achieve higher values in the error detection and correction rates. The increment in these rates translates into a higher quality of the final product.

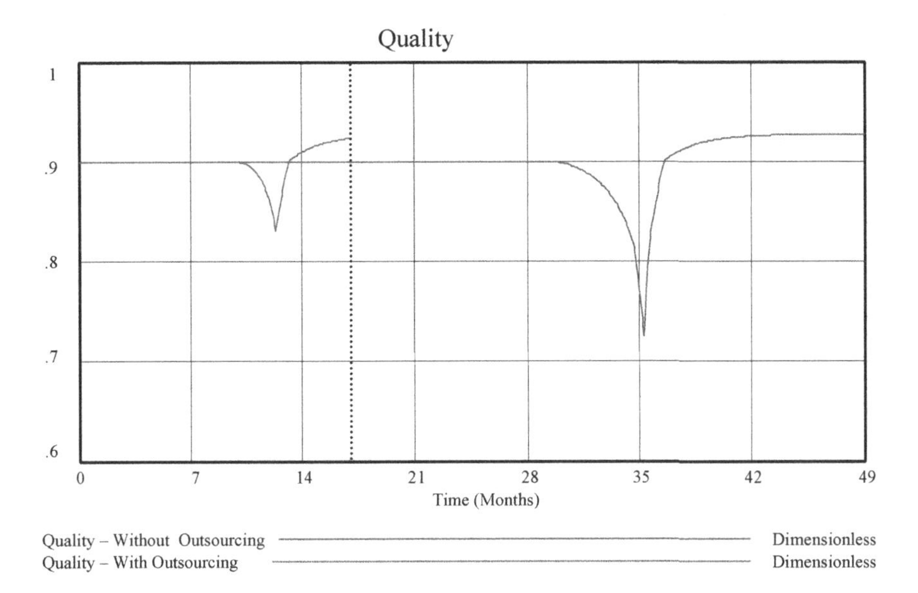

Fig. 11. Evolution of the variable *Quality*.

8. Conclusions and Outlook

This chapter has focused on software process modeling and simulation together with other traditional techniques to help organizations improve their maturity level according to CMM. There is an important factor that plays a decisive role in the achievement of this improvement. This factor is the knowledge that the organization has of its processes. It is in this field where the modeling and simulation approach can offer important advantages. The first one lies in the actual model building process. A model is a mathematical abstraction of a real system. To effectively build a simulation model, it is necessary to define *what* it is intended to model, define its *scope* and identify the rules that govern its behavior. These three activities share a common requirement: knowledge about the real system. Without knowledge, there is no information and, therefore, models. According to CMM, without knowledge, it is not possible to define the software process and therefore, to improve the maturity level. Therefore, as far as process maturity level is concerned, knowledge and process improvement go hand in hand.

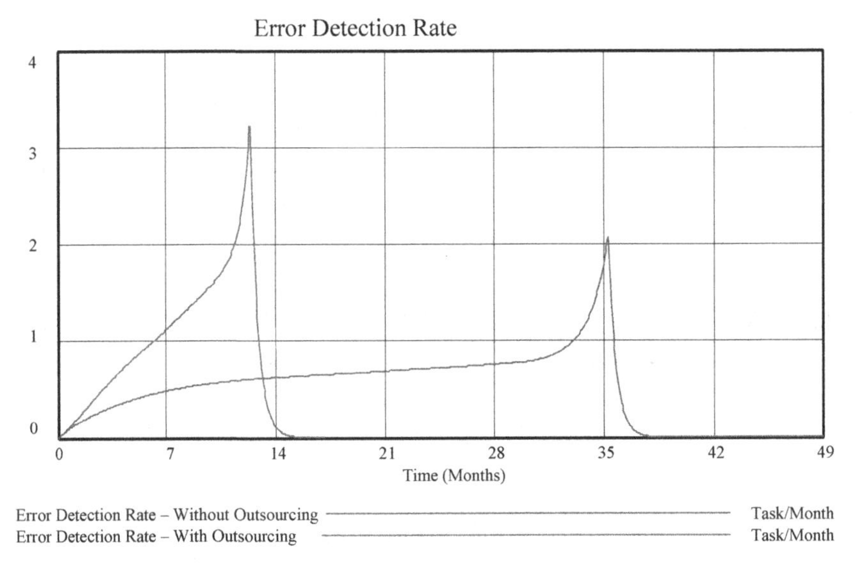

Fig. 12. Evolution of the variable *Error detection rate*.

On the other hand, simulation has always been considered as a powerful tool in the decision-making area. In this chapter, simulation has been proposed not only as a tool to help in the decision-making process, but as a factor that helps to design and evaluate process improvements. It also promotes simulation modeling and modular model building as an approach to automatically trigger the set of metrics that need to be collected, since each new dynamic module developed requires its own set of initial parameters. These initial parameters required to initialize each dynamic module should form part of the metrics collection program carried out within the organization. In addition, this new data is not only used in the simulation runs, but also to increase the level of knowledge that the organization has of its processes.

As an example of how to integrate traditional software engineering methods with software process simulation modeling, a dynamic integrated framework for software process improvement has been introduced. This framework can build dynamic software process models by means of model abstraction, module construction and reuse. These models can then be used to design and evaluate software process improvements such as analyzing the impact of the size of the technical staff on the main four variables (time, cost, quality, and overall workforce) at a level 1 organization[40] or evaluating the impact of carrying out formal inspection activities in level 3 organizations[41].

Currently, the software process modeling and simulation community is working on the application of this technique to the latest aspects of the software engineering field, such as updating the framework to work according to the CMMi[42]. Some remarkable applications are: web-based open software development[43], open source software evolution[44], extreme programming[45] and COTS-based development[46].

Acknowledgments

This work was supported by the Interministerial Commission of Science and Technology, (CICYT, Spain) through grant TIN 2004-06689-C03-03.

References

1. M.C. Paulk, B. Curtis, M.B. Chrissis and Weber, C.V. *Capability Maturity Model for Software, Version 1.1*, Software Engineering Institute, Technical Report CMU/SEI-93-TR-24. Software Engineering Institute, Carnegie Mellon University, Pittsburgh, PA. February 1993.
2. International Standard Organization. ISO 9001. Quality Systems – Model for Quality Assurance in Design/Development, Production, Installation, and Services, 1987.
3. International Standard Organization. ISO 9000-3. Guidelines for the Application of ISO9001 to the Development, Supply, and Maintenance of Software, 1991.
4. A. Carleton, R.E. Park, W.B. Goethert, W.A. Florac, E.K. Bailey and Pfleeger, S.L. *Software Measurement for DoD Systems: Recommendations for Initial Core*

Measures. Technical Report CMU/SEI-92-TR-19. Software Engineering Institute, Carnegie Mellon University, Pittsburgh, PA. 1992.

5. A.M. Christie. *Simulation – An Enabling Technology in Software Engineering.* http://www.sei.cmu.edu/publications/articles/christie-apr1999/christie-apr1999html
6. M.C. Kellner, R.J. Madachy and Raffo, D.M. *Software Process Simulation Modeling. Why? What? How?* Journal of Systems and Software, 46 (1999) 91-105.
7. High Performance Systems, Inc. 45 Lyme Road. Hannover, NH, 03755. http://www.hps-inc.com/edu/stella/stella.htm
8. PowerSim Corporation. 1175 Hendon Parkway, Suite 600, Hendon, VA, 20170. http://www.powersim.com/default_home.asp
9. Ventana Systems, Inc. 60 Jacob Gates Road, Harvard, MA 01451. http://www.vensim.com
10. Proceedings of the 5th International Workshop on Software Process Simulation and Modeling. ProSim 2004. May 24- 25, 2004. Edinburgh, Scotland UK.
11. Proceedings of the 6th International Workshop on Software Process Simulation and Modeling. ProSim 2005. May 14-15, 2005. Saint Louis, MO, USA.
12. D.M. Raffo. *Modeling Software Processes Quantitatively and Assessing the Impact of Potential Process Changes on Process Performance*. Ph.D. Dissertation. Graduate School of Industrial Administration, Carnegie Mellon University, Pittsburgh, MA. 1996.
13. M. Kellner. *Software Process Modelling Support for Management Planning and Control*. Proceedings of the First International Conference on the Software Process. Redondo Beach, California. IEEE Computer Society Press, Los Alamitos, CA (1991) 8-28.
14. G.A. Hansen. *Simulating Software Development Processes*. IEEE Computer, January 1996, 73-77.
15. P.M. Senge. *The Fifth Discipline*. Currency, 1st. Edition, 1994.
16. T. Abdel-Hamid and Madnick, S. *Software Project Dynamics: an Integrated Approach*. Prentice-Hall, 1991.
17. D. Pfhal and Lebsant, K. Integration of System Dynamics Modelling with Descriptive Process Modelling and Goal-Oriented Measurement. Journal of Systems and Software, 46 (1999), 135-150.
18. S. Burke. *Radical Improvements Require Radical Actions: Simulating a High Maturity Organization.* Technical Report CMU/SEI-96-TR-025, ESC-TR-96-024. Software Engineering Institute, Carnegie Mellon University Pittsburgh, PA, 1996.
19. P. Wernick, and Hall, T. Simulating Global Software Evolution Processes by Combining Simple Models: An Initial Study. Software Process: Improvement and Practice, 7 (2002) 113-126.
20. F.P. Brooks, Jr. *The Mythical Man-Month. Essays on Software Engineering.* 20th. Anniversary Edition. Addison Wesley – Pearson Education, 1995.
21. C. N. Parkinson. *Parkinson's Law: The Pursuit of Progress*, London, John Murray 1958.

22. J.W. Forrester. *Industrial Dynamics*. Walthan, MA: Pegasus Communications, 1961.
23. C.W. Kirkwood. System Dynamics Methods: A Quick Introduction. Technical Report. College of Business, Arizona State University, Tempe, 1998.
24. L.B. Sweeny and J.D. Sterman. *Bathtub Dynamics: Initial Results of a Systems Thinking Inventory*. System Dynamics Review 16 (4): 249-286.
25. I.J. Martinez and Richardson, G.P. *Best Practices in System Dynamics Modeling*. Proceedings of the 19th International Conference of the System Dynamics Society. Atlanta, GA USA, 2001.
26. J.W. Forrester and Senge, P.M. *Tests for Building Confidence in System Dynamics Models* In Legasto, A.A. Jr., Forrester, J.W. and Lyneis, T.M. (eds.). *System Dynamics*. New York Elsevier North-Holland, 1980, 209- 228.
27. J.D. Tvedt. *An Extensible Model for Evaluating the Impact of Process Improvements on Software Development Cycle Time*, Ph.D. Dissertation, Arizona State University, 1996.
28. L.H. Putnam and Meyers, W. Measures for Excellence: reliable software, on time, within budget. Prentice Hall, 1991.
29. B. Boehm. *Software Engineering Economics*. Prentice Hall, Inc., 1981.
30. B. Boehm, E. Horowitz, R.J. Madachy, D. Reifer, B.K. Clark, B. Steece, A.W. Brown, S. Chulani and Abts, C. *Software Cost Estimation with COCOMO II*. Prentice Hall, Inc., 2000.
31. T.L. Woodings. *A Taxonomy of Software Metrics*. Software Process Improvement Network (SPIN), 1995.
32. S.T. Knox. *Modeling the Cost of Software Quality* Digital Technical Journal, Vol, 5, No. 4 (fall 1993), 9-16.
33. D. Houston and Keats, JB. *Cost of Software Quality: a Means of Promoting Software Process Improvement*. Quality Engineering, 10(3), 563-573, 1998.
34. Q.W. Fleming and Koppleman, J.M. *Earned Value Project Management*, 2nd Edition. Newton Square, Project Management Institute, 1999.
35. W. Lipke and Jennin, M. Software Project Planning, Statistics and Earned Value. Crosstalk, December 2000.
36. I. Ramos, J.C Riquelme and Aroba, J. *Improvement in the Decision Making in Software Projects*. Miranda, P., B. Sharp, A. Pakstas, and J. Gilipe (eds.) Proceedings of the 3rd International Conference on Enterprise Information Systems (ICEIS 2001) (on CD-ROM).
37. J.R. Quinlan. *C4.5: Programs for machine learning*. Morgan Kauffman, 1993.
38. J.C. Riquelme, J.S. Aguilar and M. Toro M. *Discovering Hierarchical Decision Rules with Evolutive Algorithms in Supervised Learning*. International Journal of Computer, Systems and Signals 1(1): 73-84, 2000.
39. R. Agrawal. Mining quantitative association rules in large relational tables, ACM SIGMOD Record, v.25 n.2, 1-12, June 1996.

40. M. Ruiz, I. Ramos and Toro, M.*A. Dynamic Integrated Framework for Software Process Improvement*. Software Quality Journal (10): 181-194, 2002.
41. M. Ruiz, I. Ramos and Toro, M. *Integrating Dynamic Models for CMM-Based Software Process Improvement*. Oivo, M., and S. Komi-Sirviö (eds.) Proceedings of the 4th International Conference PROFES 2002. LNCS 2559. Rovaniemi (Finland), 63-77.
42. M.B. Chrissis, M. Konrad and Shrum, S. CMMi: Guidelines for Integration and Product Improvement. SEI Series in Software Engineering. Addison-Wesley, 2003.
43. C. Jensen and Scacchi, W. *Process Modeling Across the Web Information Infrastructure*. Proceedings of the 5th International Workshop on Software Process Simulation ands Modeling. ProSim 2004. May 24- 25, 2004, 40-49. Edinburgh, Scotland UK.
44. N. Smith, A. Capilupi and Ramil, J.F. *Qualitative Analysis and Simulation of Open Source Software Evolution.* Proceedings of the 5th International Workshop on Software Process Simulation ands Modeling. ProSim 2004. May 24- 25, 2004, 103-112. Edinburgh, Scotland UK.
45. A. Cau, G. Concas, M. Melis and Turnu, I. *Evaluate XP Effectiveness Using Simulation Modeling.* H. Baumeister, M. Marchesi and M.Holcome (eds.). Proceedings of the Extreme Programming and Agile Processes in Software Engineering. XP 2005. LNCS 3556. June 18-23, 2005, 48-56. Sheffield, UK.
46. M. Ruiz, I. Ramos and Toro, M. *Using Dynamic Modeling and Simulation to Improve the COTS Software Process.* In F. Bomarious and H. Iida (eds.): PROFES 2004, LNCS 3009, 568-581, 2004.

Chapter 3

SOFTWARE PROCESS SIMULATION WITH SYSTEM DYNAMICS — A TOOL FOR LEARNING AND DECISION SUPPORT

Dietmar Pfahl[1], Günther Ruhe[2], Karl Lebsanft[3], Michael Stupperich[4]

[1]*Fraunhofer Institute for Experimental Software Engineering (IESE)*
Sauerwiesen 6, D-67661 Kaiserslautern, Germany
E-mail: pfahl@iese.fraunhofer.de

[2]*University of Calgary*
2500 University Drive NW, Calgary, Alberta, Canada T2N 1N4
E-mail: ruhe@ucalgary.ca

[3]*Siemens AG, Dept. CT SE 3,*
Otto-Hahn-Ring 6, D-81730 München, Germany
E-mail: Karl.Lebsanft@siemens.com

[4]*DaimlerChrysler AG, Research and Technology,*
P.O.Box 2360, D-89013 Ulm, Germany
E-mail: Michael.Stupperich@daimlerchrysler.com

The chapter provides an overview of some issues in learning and decision support within the scope of software process management. More specifically, the existing work done in the field of software process simulation is presented, and the application of software process simulation models for the purpose of management learning and decision support is motivated and described. Examples of simulation modeling applications in the automotive industry are reported to illustrate how process simulation can become a useful management tool for the exploration and selection of alternatives during project planning, project performance, and process improvement. The chapter concludes with a discussion of limitations and risks, and proposes future work that needs to be done in order to increase acceptance and dissemination of software process simulation in the software industry.

1. Learning and Decision Support in the Context of Software Process Management

Software development is a highly dynamic field that heavily relies on the experience of experts when it comes to learning, applying, evaluating, disseminating and improving its methods, tools, and techniques. The experience factory has proven to be a successful approach to organization-wide, systematic improvement in software organizations.[7] It helps software organizations institutionalize their collective learning by establishing a separate organizational element that supports reuse of project experience by developing, updating, and delivering experience packages. The experience factory concepts lay the managerial and structural foundations for organizational learning and are the key to making individual competence development sustained.[5] The main idea of experience-based learning and improvement is to accumulate, structure, organize, and provide any useful piece of information being reused in forthcoming problem situations. Reuse of know-how is essentially supported by the case-based reasoning methodology.[4] However, software development and evolution typically is large in size, of huge complexity, with a large set of dynamically changing problem parameters. In this situation, reuse of experience alone is a useful, but insufficient approach to enable proactive decision analysis.[25] Diversity of project and problem situations on the one hand, and costs and availability of knowledge and information organized in a non-trivial experience (or case) base on the other hand, are further arguments to qualify decision-making. In addition to retrieving static knowledge, decisions are supported by aggregating, structuring, computing and interpreting existing units of knowledge and information.

The idea of offering decision support (DS) always arises when decisions have to be made in complex, uncertain and/or dynamic environments. The process of software development and evolution is an ambitious undertaking. In software process management, many decisions have to be made concerning processes, products, tools, methods and techniques. From a decision-making perspective, all these questions are confronted by different objectives and constraints, a huge number of variables under dynamically changing requirements, processes, actors, stakeholders, tools and techniques. Very often, this is combined with incomplete, fuzzy or inconsistent information about all the involved artifacts, as well as with difficulties regarding the decision space and environment.

Typically, a real DS system is focused on a relatively narrow problem domain. There are three kinds of existing research contributions to Software Engineering Decision Support (SEDS). Firstly, research that explicitly mentions an effort to provide decision support in a focused area of the software life cycle. Examples include decision support for reliability planning[23], inspection process management[13], multi-project management[12], or software release planning[24]. Secondly, research that contributes to decision support, although not explicitly stated as such. Basically, most results from empirical software engineering, software measurement, and software process simulation can be seen to belong to this category. Thirdly, there is an increasing effort to develop intelligent decision support systems in many other fields such as health care, transportation, finance or defense. On recent developments in this field, we refer to a paper by Shim et al.[28]

What are the expectations and requirements for systems offering SEDS? We define a set of "idealized" requirements on support systems that combine the intellectual resources of individuals and organizations with the capabilities of the computer to improve decision-making effectiveness, efficiency and transparency.[25] Depending on the actual problem topic and the usage scenario of the DS system (on-line versus off-line support, individual versus group-based decision support, routine versus tactical versus strategic support), different aspects will become more important than others.

- **(A1) Knowledge, model and experience management** of the existing body of knowledge in the problem area (in the respective organization).
- **(A2) Integration** into existing organizational information systems (e.g., ERP systems).

- **(A3) Process orientation** of decision support, i.e., consider the process of how decisions are made, and how they impact development and business processes.
- **(A4) Process modeling and simulation component** to plan, describe, monitor, control and simulate ("what-if" analysis) the underlying processes and to track changes in their parameters and dependencies.
- **(A5) Negotiation component** to evolutionarily find and understand compromises.
- **(A6) Presentation and explanation component** to present and explain generated knowledge and alternative solutions in various customized ways to increase transparency.
- **(A7) Analysis and decision component** consisting of a portfolio of methods and techniques to evaluate and prioritize generated solution alternatives and to find trade-offs between the conflicting objectives and stakeholder interests.
- **(A8) Intelligence component** to support knowledge retrieval, knowledge discovery and approximate reasoning.
- **(A9) Group facilities** to support electronic communication, scheduling, document sharing, and access to expert opinions.

The principal architecture of a SEDS system is shown in Figure 1-1.[25] Real-world decisions on planning, development or evolution processes in software engineering are made by humans. All support is provided via a graphical user interface. Experts and their human intelligence are integrated via group support facilities. The intelligence of the support is based on a comprehensive model, knowledge and experience. The more reliable and valid the models are, the more likely we are to get good support. The accompanying suite of components interacts with the model, knowledge and experience base. The suite encompasses tools for modeling, simulation, as well as decision analysis. Furthermore, intelligent components for reasoning, retrieval and navigation are added to increase efficiency and effectiveness of the support.

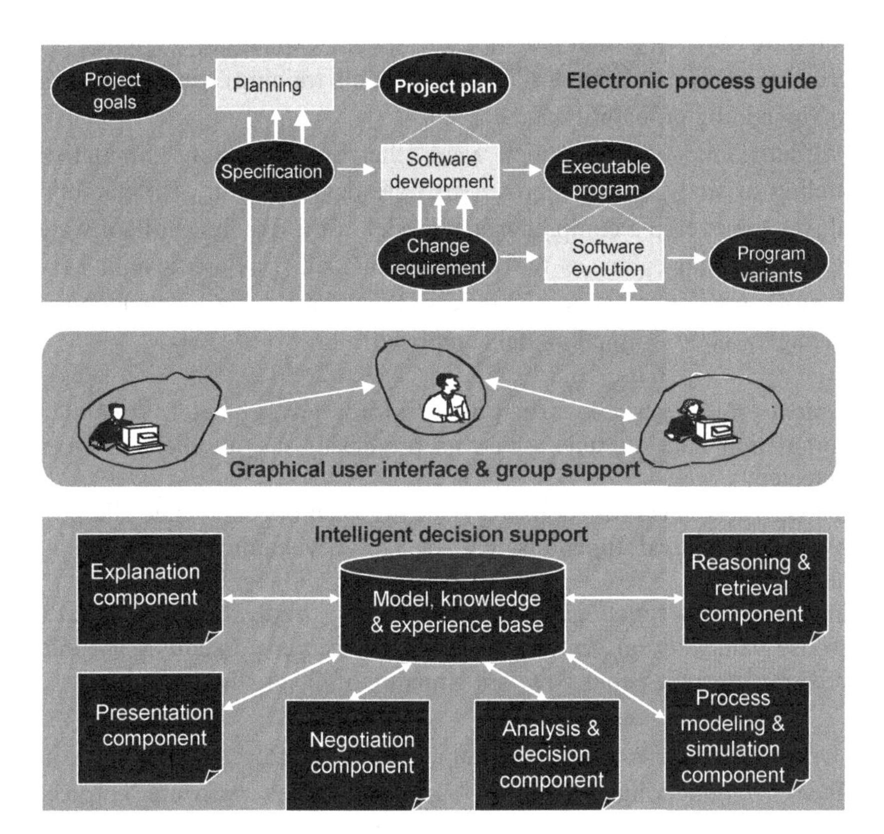

Fig. 1-1. Principal Architecture of a Software Engineering Decision Support System

In the remainder of this chapter, we mainly focus on aspect (A4) – "Process modeling and simulation component" – and explain in more detail how the process simulation component of a SEDS system can be developed and applied to support learning and decision-making.

2. Process Simulation as a Tool for Learning and Decision Support

Several authors have stressed the potential of simulation as an analysis, learning, and decision-support tool for software managers.[3,9,10,18,31] Typically, process simulation in the context of management learning and decision-support focuses on planning and controlling (re-planning) projects, analyzing past project behavior (post-mortem), exploring causes for failure and potential for improvement, explaining complex or interesting dynamic behavior to others (incl. training), and analyzing risks (by adding probability and cost).

Abdel-Hamid and Madnick were the first to apply simulation modeling to analyze and improve software development.[1] The authors' goal was to develop a comprehensive model of the dynamics of software development that enhances the understanding of, provides insight into, and makes predictions about the process by which software development is managed. Their original model comprised:

- typical project variables, such as workforce level, budget, scheduled completion date, number of errors produced, and number of errors detected;
- managerial-related functions, e.g. staffing, planning, and controlling;
- production-related functions, e.g. design, development, verification, and rework;
- human-related functions, e.g. productivity, motivation, error rate, whose values are affected by the project's perceived status (schedule pressure) and the penalty-reward structure of the organization.

During the 1990s many new applications of process simulation in software engineering, not only restricted to software project management, were published. For example, based on an exhaustive literature survey, covering the 1991-2001 period, the following application domains of process simulation in software engineering have been identified[16]: software project management, concurrent software engineering, software requirements engineering, software process improvement, software verification and validation, software reliability management, software maintenance, software evolution, software outsourcing, and software engineering training.

In an attempt to systematize the variety of applications of process simulation in the field of software engineering, Kellner et al. suggested a two-dimensional taxonomy that allows for classification according to purpose and scope.[22] The purpose dimension distinguishes the categories "strategic management", "planning", "control and operational management", "process improvement and technology adoption", "understanding", and "training and learning". The scope dimension distinguishes the following categories:

- a portion of the life cycle, e.g., design phase, code inspection, some or all of testing, requirements management;
- a development project, i.e., single product development life cycle;
- multiple, concurrent projects, e.g., across a department or division;
- long-term product evolution, i.e., multiple, successive releases of a single product;
- long-term organization, e.g., strategic organizational considerations spanning successive releases of multiple products over a substantial time period.

Extending the above-mentioned survey[16], Table 2-1 shows the distribution of software process simulation applications using Kellner's taxonomy. In total, more than 250 papers related to the topic of software process simulation published between 1987 and 2004 were identified and further analyzed. Sources were the following journals and conference or workshop proceedings: ACM Communications, IEEE Transactions on Software Engineering, Information and Software Technology, Software Process Improvement and Practice, Software Quality Journal, The Journal of Systems and Software, proceedings of SEKE, ICSE, ProSim, and PROFES. Table 2-1 classifies 81 papers presenting simulation models and related applications. Note that a paper could be classified more than once. For example, when the model's scope was "development project" and its intended usage was multi-purpose, say for planning, controlling, and improving, it was classified three times. Most of the simulation modeling projects while focusing on complete development projects or portions of it aimed at planning, improvement and technology adoption, and understanding.

3. Guidance for System Dynamics Process Simulation Modeling

As an initial input to those who wish to learn more about the development of process simulation models, we introduce IMMoS

(Integrated Measurement, Modeling and Simulation), a method for goal-oriented development of System Dynamics (SD) simulation models.

Table 2-1. Classification of process simulation applications

Scope / Purpose	Portion of life cycle	Development project	Multiple, concurrent projects	Long-term product evolution	Long-term organization
Strategic management		6	2	2	6
Planning	4	20	3	1	2
Control and operational management	2	6	2		
Process improvement and technology adoption	6	18	1	1	
Understanding	6	16	3	3	2
Training and learning		11			1

SD is a very comprehensive and powerful simulation modeling paradigm. SD models are able to capture both static and dynamic aspects of reality. In addition, they provide two fully consistent representation layers: a qualitative graphical representation layer, the so-called flow graph, and a quantitative representation layer consisting of a set of well-defined mathematical equations. The graphical representation layer is useful for model building and human understanding of model contents, while the mathematical representation layer is useful for running simulations and conducting experiments with the model in a virtual laboratory setting. The basic ingredients of flow graphs are shown in Figure 3-1.

SD model equations are separated into two groups: level equations and rate equations. Level equations describe the state of the system. They accumulate (or integrate) the results of action in the systems, an action being always materialized by flows in transit. The derivative of a level, or equivalently the rapidity at which it is changing, depends on its input and output flows. The rates are what change the values of levels. Their equations state how the available information is used in order to generate actions. Finally, constants are used to parameterize SD models, while auxiliary variables are used to improve the readability of model equations by storing intermediate calculation results.

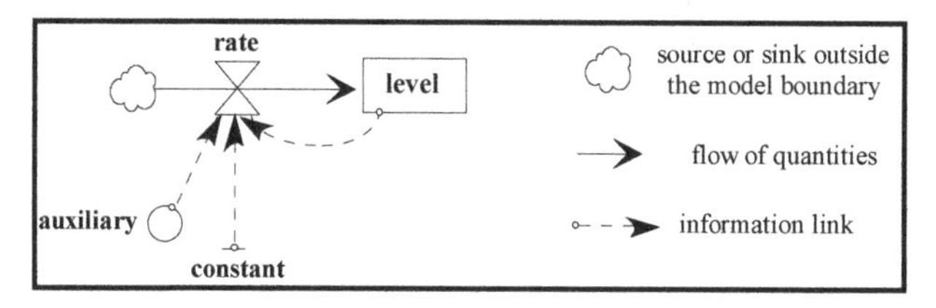

Fig. 3-1. Schematic conventions of flow graphs

Although several authors, starting with the seminal work done by Forrester[11], have published phase models and process descriptions[l], there is no detailed guidance in the form of a process model that defines entry and exit criteria for each SD modeling activity, enforces precise problem definition, helps to identify stakeholders based on an associated role model, defines the product flow based on an associated product model, provides templates and checklists, and offers hints on when and how to reuse information from other software modeling activities.

In order to resolve these shortcomings, and thus improve the efficiency and effectiveness of SD model development in the software engineering context, a comprehensive methodology for Integrated Measurement, Modeling, and Simulation (IMMoS) was developed.[19] The IMMoS methodology consists of four components:

- C1 (Process Guidance) provides a model of the SD development life cycle with associated role model, product model, and process model.
- C2 (Goal Definition) provides a SD modeling goal definition taxonomy specifying five dimensions (role, scope, purpose, dynamic focus, and environment) that capture the problem definition in the early stage of SD model development.
- C3 (Model Integration) describes how static software engineering models like process models (descriptive and prescriptive) and quantitative models (e.g., cost, quality and resource models) are integrated with SD models.

---

[l] A summary of these proposals can be found in the doctoral thesis by Pfahl.[16]

- C4 (Method Integration) describes how SD model development relates to process modeling and goal-oriented measurement following the Goal/Question/Metric (GQM) paradigm.[8] Particular focus is put on the integration of SD model development with GQM, enhancing the established GQM method towards "Dynamic GQM".[20]

In the following, we briefly present component C1 (Process Guidance). Process guidance is facilitated through a set of models that support the SD model developer: Role Model, Phase Model, Product Model, and Process Model.

3.1. IMMoS Role Model

The IMMoS Role Model defines the minimal set of roles that are typically involved in SD simulation modeling projects: Customer (C), User (U), Developer (D), Software Engineering Subject Matter Expert (E), Facilitator (F), and Moderator (M).

Role C (Customer) represents the sponsor of the SD modeling project. For the SD modeling project to be successful it is important that C knows about the cost and benefit of developing and using SD models. This includes a basic understanding of typical application areas of SD models. C is responsible for the identification of potential SD model users, and of providing the human resources (i.e., Software Engineering Subject Matter Experts) for the SD model development and maintenance task.

Role U (User), i.e., the future user of the SD model in the software organization, is responsible for providing the necessary information for SD modeling goal definition. In addition, U participates in all phases of the SD modeling life cycle, particularly during verification and validation activities, and during the definition of the SD model user interface (when desired). During SD model application, U triggers enhancements of the existing model, e.g., re-calibration of the model parameters due to changes in the real world.

Role D (Developer) is responsible for technically sound SD model development. In order to fulfill this task, the following skills are needed:

- Sufficient theoretical and practical knowledge about the SD modeling approach gained through training, relevant literature, and, ideally, active participation in previous SD modeling projects.
- Sufficient knowledge about at least one SD modeling tool.
- Sufficient communication skills and ability to apply knowledge elicitation, moderation, and presentation techniques.
- Sufficient knowledge about measurement (i.e., GQM) and process modeling.
- Basic knowledge about the organizational and technical characteristics of the environment in which the SD simulation modeling project takes place is useful.

Role E (Software Engineering Subject Matter Expert) is responsible for providing the relevant SE information needed for SD model building. This includes managerial and technological information about how software is developed (processes, methods, techniques, tools, plans, measurement data, etc.) in the organization

Role F (Facilitator) helps with establishing contacts, and planning and arranging meetings. The responsibility of F is strictly limited to technical support during a SD modeling project. This role is often taken over by the same person that assumes role C, U, or even D – if the person assuming D is familiar with the customer organization.

Role M (Moderator) is responsible for preparing and leading workshops and meetings of D with three or more subject matter experts (cf. role E).

3.2. IMMoS Phase and Product Models

The IMMoS Phase Model structures the SD modeling life cycle into four phases. Phase 0 prepares the actual model building activities, while phases 1 to 3 represent the (iterative) life cycle that SD models typically follow.

Phase 0 (Pre-Study) focuses on the definition of prospective SD model users, identification of software engineering subject matter experts that can be approached by the SD model developer during the modeling activities, and the specification of the SD modeling goal. If no SD model user can be identified or no precise SD modeling goal definition can be achieved, the modeling activity should be stopped. The following artifacts are produced during phase 0: Agreement, Customer Sheet, Management Briefing Materials, Management Briefing Minutes, Goal Definition, Project Plan, Project Log file, and Development Contract.

In phase 1 (Initial Model Development), an initial SD model is developed that is able to reproduce the reference mode. The reference mode is an explicit description of the (problematic) dynamic behavior of one or more system parameters observed in reality. It acts as a catalyst in the transition from general speculation about a problem to an initial model, and it captures the dynamics of the tackled problem, i.e., behavior patterns and related time horizon. The following artifacts are produced during phase 1: Technical Briefing Materials, Technical Briefing Minutes, Development Workshop Minutes, Dynamic Hypotheses Definition (consisting of Reference Mode and Base Mechanisms), Causal Diagram, Verification Report 1, Initial SD Model (consisting of Initial Flow Graph, Initial SD Model Equations, Initial SD Model User Interface), Verification Report 2, and Validation Report 1.

In phase 2 (Model Enhancement), the initial SD model is enhanced such that it can be used for problem solving. It might be the case that the SD model user is only interested in a singular problem solution, e.g., when the goal is to evaluate alternative improvement suggestions. In this case, the modeling activities would stop at the end of phase 2. The following artifact is produced during phase 2: Enhanced SD Model (consisting of Enhanced Flow Graph, Enhanced SD Model Equations, Enhanced SD Model User Interface, and Enhanced Causal Diagram).

In phase 3 (Model Application) the enhanced SD model is applied to serve its specified purpose. If needed, the artifacts produced in phase 2 are enhanced and maintained in order to cope with a changing reality.

Detailed descriptions of the individual SD modeling artifacts can be found in more specific publications by Pfahl and Ruhe.[16,19]

3.3. IMMoS Process Model

The IMMoS Process Model provides a control-flow oriented description of the sequence of activities that should be followed in a SD model development project.

Phase 0 (Pre-Study) comprises the following seven activities: First contact, Characterization, Management briefing, Identification of model users, Problem definition, Technical feasibility check, and Planning and contract.

Phase 1 (Initial Model Development) also comprises seven activities: Technical briefing, Definition of dynamic hypotheses, Definition of the causal diagram, Review of the causal diagram (Verification 1), Implementation of the initial SD model, Review of the initial SD model (Verification 2), and Test of the initial SD model (Validation 1).

Phase 2 (Model Enhancement) comprises two activities: Enhancement of the initial SD model, and Test of the enhanced SD model (Validation 2).

Finally, phase 3 (Model Application) comprises only one activity: Application and maintenance of the SD model.

Each activity is characterized through a set of attributes, such as involved roles, entry/exit criteria and input/output products. As an example, the activity 0.5 (Problem Definition) is shown in Table 3-1. A list of all SD modeling activities can be found in a focused paper by Pfahl and Ruhe[19], while a complete description of the IMMoS Process Model has been published by Pfahl[16].

4. Applications in the Automotive Industry

In the form of examples, we present two simulation models that were developed with IMMoS in the following sections. The case studies represent applications of software process simulation models that were developed to support learning and decision-making within software organizations linked to the automotive industry.

The first model, RESIM (Requirements Simulator), was developed jointly with Siemens Corporate Technology. The aim of the RESIM project was to provide support for analyzing the potential effectiveness of improvement suggestions proposed in the context of a software process assessment in Siemens' automotive business unit. Using the IMMoS Goal Definition Template, the simulation modeling goal can be summarized as shown in Table 4-1.

Table 3-1. Example SD modeling activity description

ID	Activity 0.5
Name	Problem definition
Role	C (optional), D, U: - C: Checks whether Goal Definition is in line with business goals. - D: Supports U during problem definition. - U: Responsible for problem identification and definition.
Input	Customer Sheet If available: process models and measurement-based quantitative models.
Output	Goal Definition, Project Logfile
Entry condition	U has been identified (cf. Activity 0.4)
Exit condition	Goal Definition exists in written form or SD modeling project has been cancelled.
Description	- Identification of a problem that – if solved – would help U with his/her daily work. - Formal documentation of the problem definition (SDM Goal Definition). - Notes should be taken of all relevant information that could be used to define the dynamic hypothesis in Phase 1, e.g., first assumptions about cause-effect relationships, suggestions of potential problem solutions, relevant existing models, subject matter experts, etc. This kind of information is recorded in the Project Logfile.
Methods / Techniques	Knowledge elicitation techniques: Interview (semi-structured or unstructured) and focused discussion (goal-related)
Guidelines	- The problem definition should be well-focused and be stated in concrete terms. The Goal Definition Template should be used. - In order to be suitable for SD analysis, the problem has to deal with phenomena that show dynamic behavior. - In order to be suitable for SD analysis, the system that is going to be investigated for problem solution, has to be viewed as a feedback (or closed) loop system. This assumption implies that a change in the system structure – and not an alteration of the inputs – is in the focus of interest of the problem solution.
Materials / Tools	Goal Definition Template

The second model, PL-SIM (Process Leadership Simulator), was developed jointly with DaimlerChrysler Research. The aim of the PL-SIM project was to support the strategic software improvement planning in one of DaimlerChrysler's car divisions. The PL-SIM model provided first insights into selecting and evaluating proposed elements of strategic software process improvement programs. Using the IMMoS Goal

Definition Template, the simulation modeling goal can be summarized as shown in Table 4-2.

Table 4-1. Goal definition template for simulation model RESIM

Role	Process Consultant (Assessor)
Scope	Single Project
Dynamic Focus	Impact of software requirements volatility on software development productivity
Purpose	Understanding
Environment	Siemens Automotive (Micro Controllers)

Both simulation models, RESIM and PL-SIM, provided evidence for the effectiveness and efficiency of IMMoS, and for the suitability of SD models to serve as specialized components of SEDS systems. In particular with regards to efficiency, using IMMoS significantly saved simulation modeling effort and shortened model development time.[16]

Table 4-2. Goal definition template for simulation model PL-SIM

Role	Software Engineering Process Group
Scope	Software Organization
Dynamic Focus	Impact of improvement measures on process leadership
Purpose	Understanding
Environment	DaimlerChrysler Automotive

The following two sections describe in more detail how RESIM and PL-SIM were applied for learning and decision support in two different software organizations linked to the automotive industry.

4.1. Simulation in Support of Software Process Assessment

The SD model RESIM was developed jointly by Fraunhofer IESE and Siemens Corporate Technology (Siemens CT). The purpose of this simulation model was a) to demonstrate the impact of unstable software requirements on project duration and effort, and b) to analyze how much effort should be invested in stabilizing software requirements in order to achieve optimal cost effectiveness.

The starting point for developing RESIM was a CMM-compatible software process assessment[14,30], which Siemens CT had conducted within a Siemens Business Unit (Siemens BU). Usually, the main result of a software process assessment is a list of suggested changes to the software processes. In this case, the assessors' observations indicated that the software development activities were strongly affected by software requirement volatility. Moreover, due to the type of products developed by Siemens BU, i.e. products consisting of hardware (e.g. micro-controllers) and embedded software, the definition of software requirements was under direct control of systems engineering, and thus the software department did not hold full responsibility. During the assessment, the assessors observed that many system requirements that had already been addressed by software development were changed by the customer, or replaced by new requirements defined by systems engineering late in the project. In addition, there were many cases where system requirements that originally had been passed to software development eventually were realized by hardware, and vice versa. Based on these observations, the assessors expected that improvement suggestions that exclusively focused on software development processes (e.g., introduction of software design or code inspections) would not help stabilize software requirements. Since the software department that had ordered the process assessment primarily requested improvement suggestions that could be implemented under their responsibility, there was a need to find means that helped convince decision makers that first systems engineering had to be improved before improvements in software development could become effective. Hence the decision was made to develop a simulation model, i.e. RESIM, which could help clarify the situation and investigate the cost-effectiveness of improvements in systems engineering with regards to software development.

Following the IMMoS approach, the model building process was highly iterative. Thirteen increments were needed to come up with a base model that was able to capture the software development behavior mode of interest, and which contained all relevant factors governing observed project behavior. After two additional iterations, the simulation model was ready to be used for its defined purpose.

Besides the definition of the model boundaries and model granularity, the most important design decisions were related a) to the typically observed behavior patterns ("reference mode") of development projects at Siemens BU which the model was to be able to reproduce through simulation, and b) to the assumptions about the most significant cause-effect relationships ("base mechanisms") governing the observed project dynamics.

The reference mode represents the dynamics of product evolution and requirement generation. The software product is developed in three successive periods of approximately the same length. One increment is developed during each period. Each product increment implements certain types of requirements.

At the beginning of each increment development period, there is a known fixed set of requirements to start with. During the development of an increment, new requirements are received from the customer. Typically, the number of new requirements exhibits a ceiling effect.

To build the SD model it was necessary to identify the most important causal relationships that are believed to generate the typical project behavior. The starting point for this modeling step was the assumption that the stability of software (SW) requirements definition is a measure of systems engineering (SE) quality, and that systems engineering quality can be increased if effort is invested in related improvement actions. Based on the insights that the Siemens CT experts gained during process assessment, the following base mechanisms were identified:

- The more effort is spent on SE improvement, the better the quality of SE is: [SE effort + → SE quality +]
- The better the quality of SE is, the higher the stability of the SW requirements is: [SE quality + → stability of SW requirements +]

- The higher the stability of SW requirements is, the smaller the number of implemented SW requirements that have to be replaced or modified is: [stability of SW requirements + → replacement of implemented SW requirements -]
- The more requirements that have already been implemented are replaced by new requirements, the larger the total number of requirements to be implemented, and thus the time needed to complete the project is: [replacement of implemented SW requirements + → total number of SW requirements to implement + → SW project duration +]
- The more (excess) time is needed to complete the project, the bigger time pressure is: [SW project duration + → time pressure +]
- The bigger the time pressure is, the more (additional) manpower will be allocated to the project: [time pressure + → manpower +]
- The more (additional) manpower is allocated, the bigger the average development productivity will be: [manpower + → development rate (per time unit) +]
- The more requirements that have already been implemented are replaced by new requirements, the more iterations (so-called I-cycles) have to be conducted: [replacement of implemented SW requirements + → number of I-cycles +]
- The more I-cycles are conducted, the smaller the average development productivity of the related increment is: [number of I-cycles + → development rate (per time unit) -]

To better understand the key dynamics of the system to be modeled, the individual causal relationships can be linked together in a causal diagram (cf. Figure 4-1).

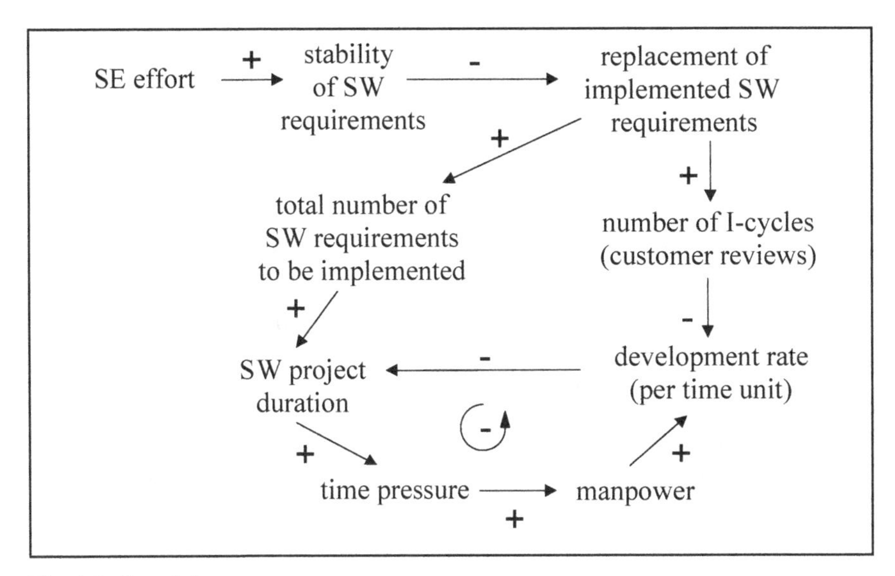

Fig. 4-1. Causal diagram

The causal diagram clearly shows that an increase of SE effort would reduce SW project duration for two reasons. Firstly, it would reduce the overall number of SW requirements that is implemented (also counting replacements or modifications of already implemented requirements). Secondly, it would reduce the number of iterations (I-cycles), and thus increase the average development rate (per time unit). Conversely, a lack of SE effort would increase SW project duration, which – in order to keep the project deadline – could only be compensated by adding manpower. This compensation mechanism is controlled through a negative feedback loop.

The simulation model was implemented in a modular way. The main module represents software development with its interface to systems engineering from which the software requirements are received. Four additional modules describe certain aspects of software development in more detail, namely: workforce allocation and adjustment, effort and cost calculations, generation of new software requirements, and co-ordination of incremental software development. Figure 4-2 shows how the five modules are interrelated. In addition, the figure indicates that module 3 (effort and cost calculation) processes the variables that are needed for solving the issue under consideration, i.e. the effort allocated for systems engineering activities (input or policy variable), and the cumulated total effort resulting from all software and systems engineering activities actually conducted (output or result variable).

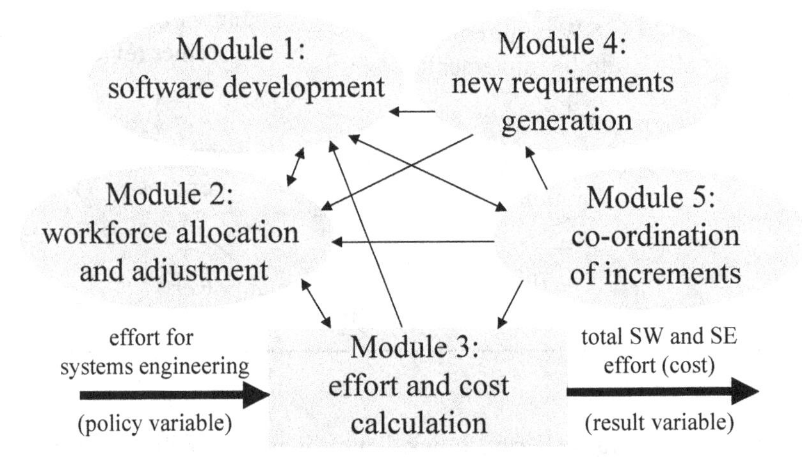

Fig. 4-2. Modular structure of the SD model with interfaces

Model validation was mainly based on plausibility checks conducted by Siemens CT experts. The most important necessary condition for model validity, i.e. the ability to reproduce the reference mode, was fulfilled.

Figure 4-3 presents the simulated patterns of SW product growth (implemented stable requirements → variable: SW_product) and growth of SW requirements that actually are contained in the final SW product (stable requirements → variable: actual_all_SW_requirements), as generated by the SD model for the baseline situation.

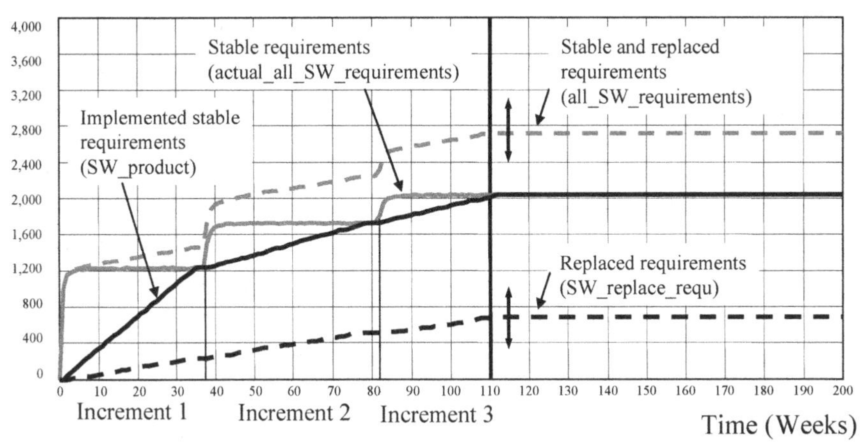

Fig. 4-3. Reproduction of the reference mode

Simulation showed that the number of replaced requirements (variable: SW_replace_requ) and thus the total number of stable and replaced requirements (variable: all_SW_requirements) can vary largely as a consequence of variation in effort invested to improve systems engineering.

Ultimately, the question that had to be answered with the help of the simulation model was: "How much effort should be invested in systems engineering in order to improve (software) requirements analysis and thus minimize the overall software development cost?" To answer this question, an equivalent mathematical minimization problem was formulated:

$$\text{total\_effort} = x + \sum_{t=1}^{T} y(t) \longrightarrow \min$$

with:

t: elapsed time (weeks)
T: project termination (weeks)
x: effort for systems engineering (person weeks)
y: weekly effort consumption for software development (person weeks / week)

The solution to this problem was found by varying the policy variable x, i.e., effort allocated for systems engineering activities, and by using the built-in optimization function of the simulation tool[2]. It turned out that an increase of the systems engineering effort share from 1.7% of the total effort (baseline situation) to 9.1% of the total effort (optimal situation) would reduce the overall cost for systems engineering and software development by more than 20%. This effect is mainly due to the stabilization of requirements, which is expressed in terms of the actual average requirements replacement per week. In the optimal case, on average only 0.08% of the currently known (and already implemented) requirements were replaced per week, adding up to a total of 29 replaced out of 1000 initially planned requirements during project performance.

Based on the simulations it was possible to demonstrate that software requirements volatility is extremely effort consuming for the software development organization and that investments in systems engineering in order to stabilize requirements definition would pay off well. The results of the model experiments have provided a twofold advantage. Firstly, a deeper understanding of the procedures for capturing and changing requirements grew up in the assessment team while discussing real life and its representation in the model. Secondly, the quantitative evaluation of the present situation and of the effect of possible changes was convincing for the Siemens BU. The model results helped a lot to broaden the view of the requirements process within software development and to start an improvement program across all the roles and organizations participating in this process.

Even if the simulation model has to be viewed as being qualitative due to the lack of precise quantitative data to which the model can be calibrated, having such a simulation model at hand makes it easier to visualize the critical project behavior and to discuss the assumptions about the cause-effect relationships that are supposed to be responsible for the generated behavior. In that sense, experts at Siemens CT felt that building the SD model was a useful exercise, and that similar models can help them in future process improvement projects with Siemens business

---

[2] Here, as in all other examples presented in this chapter, the System Dynamics simulation modeling tool Vensim was used.[29]

units. More details on the project can be found in an article by Pfahl and Lebsanft.[17]

4.2. Simulation in Support of Strategic SPI

SD model PL-SIM was developed jointly by Fraunhofer IESE and DaimlerChrysler Research and Technology. Within DaimlerChrysler (embedded) software development plays an important role due to the increasing amount of software in cars. In order to constantly improve the development of software, various activities have been initiated like experience-based process improvement[27] or the GQM method for goal-oriented measurement[26].

The specific goal of developing the SD model PL-SIM was to support DaimlerChrysler Research and Technology in conducting a study that aimed at exploring the level of process leadership (PL) achievable by one of DaimlerChrysler's car divisions (in the following abbreviated as DC/CarDiv) within the next five years by applying certain improvement measures. Mastering PL requires:

- The ability to manage complex projects with several suppliers and various customers;
- High process maturity throughout the complete (embedded) software life-cycle – including maintenance;
- Tool support of all related development processes – as formal as necessary, as flexible as possible.

Ultimately, PL is expected to yield higher product quality and help reduce warranty and fair-dealing costs.

The simulation model PL-SIM was developed in order to focus discussions and provide a quantitative basis for analyses. The underlying principles of the simulation model, and its structure and calibration are based on the input received from DaimlerChrysler Research and Technology and experts from DC/CarDiv. A crucial issue in the scope of the study was the identification and determination of factors and measures that have an impact on the level of PL.

Following the IMMoS approach, the simulation model was developed in several iterations, yielding three major model versions with more than a dozen intermediate variants.

The principles and assumptions behind the simulation model can be expressed in the form of cause-effect mechanisms. The integration of these cause-effect mechanisms results in a cause-effect structure that is believed to generate the dynamics of PL change over time. The cause-effect structure of the simulation model is presented in Figure 4-4.

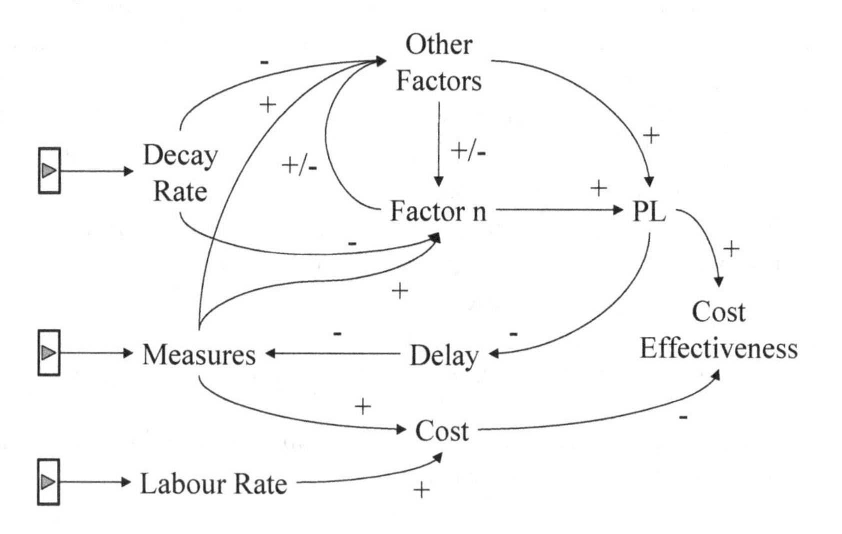

Fig. 4-4. Causal diagram

The cause-effect structure reflects the following fundamental principles behind the simulation model.

- **Principle P1:** The actually achieved level of PL depends on the level of control over the influence factors, which in turn depend on average yearly decay (Decay Rate) and measures to maintain or improve their current level of control (Measures).
    - [(Other) Factor(s) + → PL +]
    - [Decay Rate + → (Other) Factor(s) -]
    - [Measures + → (Other) Factor(s) +]
- **Principle P2:** There is a mutual dependency between influence factors. As a consequence, a change of one factor (Factor n) will induce changes of the other factors (Other Factors).

- [Factor i + → Factor j +/- depending on the type of correlation]

- **Principle P3:** The speed with which measures become effective is subject to delay (Delay). Delay, in turn, depends on the level of PL since it is easier (and thus faster) to implement measures when process leadership is high.
  - [Delay + → Measures -]
  - [PL + → Delay -]
- **Principle P4:** In order to estimate the cost effectiveness of measures taken to maintain or improve control of influence factors on PL (Cost Effectiveness), one can compare the change of PL induced by these measures with their cost (Cost). The cost of measures depends on the labor rate (Labor Rate) and amount of effort needed to perform the measures. The effort needed to perform a measure depends on the duration of the measure and the intensity with which it is conducted (not shown in the causal diagram).
  - [PL + → Cost Effectiveness +]
  - [Cost + → Cost Effectiveness -]
  - [Labor Rate + → Cost +]
  - [Measures + → Cost +]
- **Principle P5:** Because it is generally difficult to define commonly accepted absolute scales for PL, influence factors and improvement measures, only relative quantities defined on the interval [0, 1] are used in the model.[3]

The modeling approach was top-down, starting by identifying the most important factors believed to have an impact on PL, then identifying measures by which the current level of the impact factors could be improved, and eventually assessing quantitatively the starting values of PL, impact factors, and measures, as well as assessing the quantitative relationships between each of these. A more comprehensive description of the underlying modeling assumptions and experiments run with the simulation model can be found in a paper by Pfahl et al.[21]

---

[3] Strictly speaking, Principle P5 does not directly relate to the model structure. However, due to its importance as a fundamental guideline throughout the simulation modelling process, it was considered as having the same importance as a principle.

In the case of DC/CarDiv, there are six factors. One of them is organization-related, two are process-related, another two are product-related, and the last one is people-related:

- Factor OWN: Ownership (of products, processes, and human resources)
- Factor DEV: SW Development Process – applies to both DC/CarDiv and its sub-contractors
- Factor MNT: SW Maintenance Process – applies to both DC/CarDiv and its sub-contractors
- Factor FEA: Product Characteristics (Features) – customer view on the product
- Factor STR: Product Structure – development and maintenance view on the product
- Factor HRE: Human Resources

The matrix of mutual influence between factors was estimated to be as presented in Table 4-3. The matrix should be read as follows: a relative change of factor DEV by X % causes a relative change of factor OWN by 2*X % (cf. matrix cell defined by second row and third column).

Table 4-3. Mutual impact between factors that have an influence on PL

(column has impact on row)	**OWN**	**DEV**	**MNT**	**FEA**	**STR**	**HRE**
Ownership	-	2	0	1	1	1
SW Development Process	0	-	0.5	1	1	0
SW Maintenance Process	0	1	-	0.5	0.5	0
Product Characteristics	0	1	0.5	-	0	0
Product Structure	0	1	0.5	0	-	0
Human Resources	0	2	1	2	2	-

A specific list of relevant measures associated with the factors in the model was defined. The total number of measures adds up to 17. Figure 4-5 shows how many measures are associated with each factor.

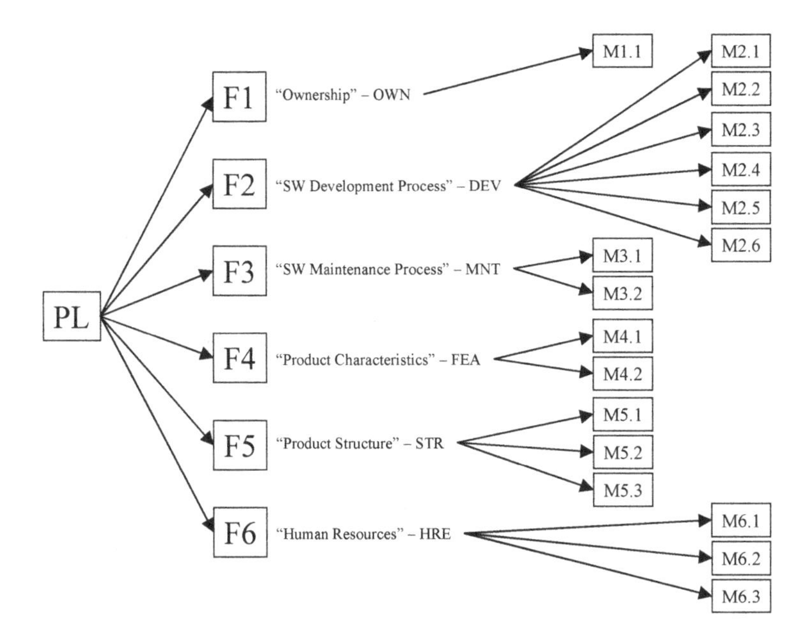

Fig. 4-5. Tree of impact factors and associated measures

As an example, Table 4-4 lists two measures related to factor FEA that are considered to be of relevance and to have improvement potential, i.e., SPP (establish strategic project planning) and PD (establish product documentation). Both SPP and PD measurement values are in [0, 1]. Only three distinct values can be assumed: 0, 0.5, and 1. The corresponding measurement rules are shown in column 3 of Table 4-4. Finally, column 4 gives some hints on possible data sources.

Several scenarios were simulated with the model. Scenarios help to analyze the impact of applying improvement actions (measures) associated with certain factors influencing PL.

Table 4-4. Example set of measures related to factor "Product Characteristics (FEA)"

Factor	Measure	Possible measurement values	Possible data sources
Product Characteristics – FEA	M4.1: Establish strategic product planning (SPP)	Value range: [0, 1] 0 = no or insufficient SPP 0.5 = SPP exists 1 = SPP exists and regular benchmarks are conducted	Competitor benchmark, customer satisfaction survey
	M4.2: Establish product documentation (PD)	Value range: [0, 1] 0 = no or insufficient PD in place 0.5 = initial PD in place 1 = comprehensive PD in place	Available/existing customer and service documentation

For example, the scenario presented below illustrates how the activation of a set of improvement actions associated with three impact factors over certain time periods with specific intensity influences the evolution of PL. Figure 4-6 indicates that measures M2.5 of factor DEV, measures M4.1 and M4.2 of factor FEA (cf. SPP and PD in Table 4-4), and measure M6.1 of factor HRE are activated at time t = 40, 30, and 50 weeks after the start of the simulation, for a time period of 200, 70, 95, and 100 weeks, respectively. In this case, each of the improvement actions is conducted in a way that its impact on the level of factor control is maximal. For example, in the cases of measures SPP, this means that a strategic product planning has been established and is benchmarked regularly.

The impact of the measures on process leadership PL (via alteration of FEA) is shown in Figure 4-7. Run4 represents the baseline, i.e., when no improvement actions are taken at all. Run3f-4 represents the behavior of the value of PL if measures M2.5, M4.1, M4.2, and M6.1 are activated as described above. It can be seen that PL can recover from decrease induced by decay of its impacting factors. In other words, the activated measures related to factors DEV, FEA, and HRE are able to compensate the continuing decay of factors OWN, MNT, and STR, areas in which no improvement activities are conducted. At around time t = 95 weeks PL starts to decrease again because the self-reinforcing decay becomes stronger, and because measures stop being activated.

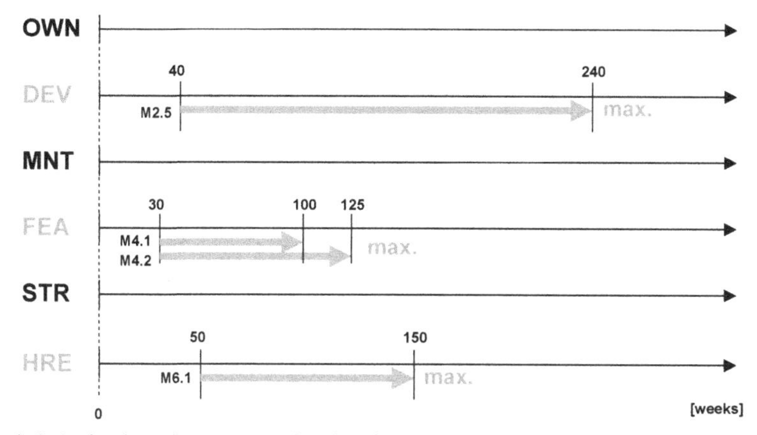

Fig. 4-6. Activation of measures related to factors DEV, FEA, and HRE

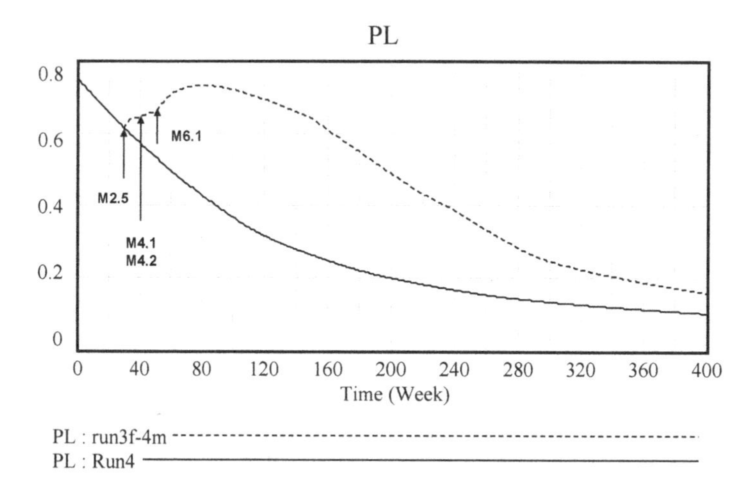

Fig. 4-7. Impact of measures M2.6, M4.1, M4.2, and M6.1 on PL (Process Leadership)

The impact of the activated measures on the individual factors is shown in Figure 4-8. Not surprisingly, a strong reaction is induced on factors DEV, FEA, and HRE. In addition, due to the mutual (positive and negative) correlation between factors (cf. Table 4-3), some reaction is also induced on factors OWN, MNT, and STR.

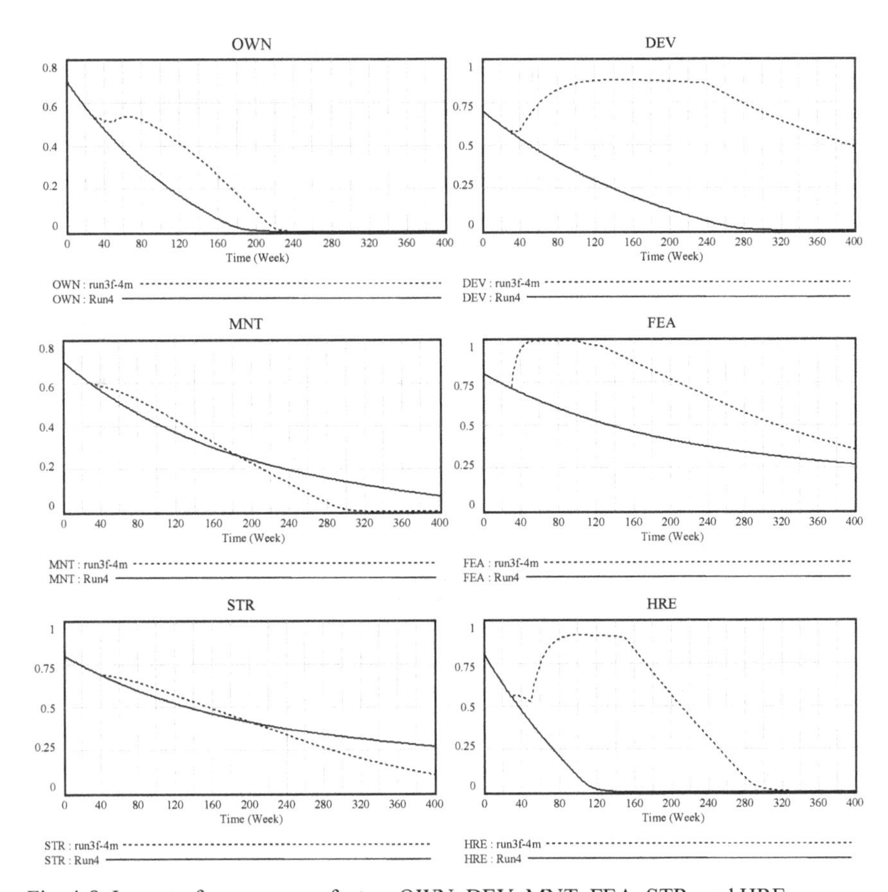

Fig. 4-8. Impact of measures on factors OWN, DEV, MNT, FEA, STR, and HRE

The simulation modeling project was generally judged as successful by experts within DaimlerChrysler, but it also had some limitations. While the simulation modeling exercise helped build well-grounded consensus regarding the appropriateness of the selected impact factors, their mutual interrelation, and the quantitative adequacy of their starting values, the degree of uncertainty about the completeness of the set of identified measures and the lack of information on how to define related measurement rules and collect data was too large to run quantitatively reliable simulations. Therefore, the main value of the model was seen as its function as a tool for management learning that can be used to generate hypotheses about the impact and long-term effects of improvement measures within strategic SPI programs.

Apart from its specific value for DaimlerChrysler, other organizations may (re-)use the generic simulation model as an easily adaptable framework that helps explore and assess their own strategic SPI programs. The current generic simulation model can easily be adapted by replacing the strategic improvement objective, i.e. PL, the set of related influence factors, and the sets of associated improvement actions (measures) with organization-specific instantiations.

5. Conclusion

Although the number of software process simulation applications is constantly growing and the efficiency of one important class of simulation modeling approaches, System Dynamics, has been improved by introducing the IMMoS method, there seems to be a persisting problem in industry with regard to accepting that software process simulation models can become a trusted component of SEDS systems and an effective tool for learning. Based on our research and experience in the field, we believe there are two main reasons behind the reluctance to use process simulation models in the software industry. First, process simulation modelers tend to develop their models from scratch. Taking into account the typical size and complexity of a software process simulation model, this is in most cases, i.e., under the given constraints, still too difficult and thus time and effort consuming. Second, there are often too high expectations with regard to required model validity. Inappropriate requirements with regard to model validity, however, in most cases increase development time and effort.

We are currently working on resolving the first issue.[6] The tendency to develop simulation models each time from scratch is mainly due to the lack of practical guidance for planning and performing comprehensive reuse of process simulation modeling artifacts (or parts of them). In order to facilitate reuse and speed up the development process, we have started to investigate possibilities for exploiting the principles of (process) design patterns[15] and agile software development[2].

Regarding the second issue, future research has to focus on methods to specify validity requirements for process simulation models that are appropriate for their intended application. For example, complex simulation models that are expected to generate accurate point estimates require a high degree of predictive validity. However, models that are mainly used as a tool for learning or thinking typically only require an appropriate degree of behavioral validity, i.e., it is sufficient that simulation runs reflect the typical dynamic behavior of key model parameters, and the typical changes of certain behavior patterns in response to alterations of simulation inputs.

If we take a more general view and go beyond the discussion of the current state-of-the-art in software process simulation modeling methodology, our experience with applying software process simulation for management learning and decision-support, as illustrated through the examples in section 4, has shown promising results which has reinforced our belief that they will play a prominent role in future SEDS systems and tools for management learning.

References

1. Abdel-Hamid, T. K., Madnick, S. E.: *Software Projects Dynamics – an Integrated Approach*. Prentice-Hall (1991).
2. Abrahamsson, P., Warsta, J., Siponen, M. T., Ronkainen, J.: New Directions on Agile Methods: A Comparative Analysis. In: *Proceedings of the 25th International Conference on Software Engineering*, IEEE, Portland, Oregon, USA (2003), 244-254.
3. Acuña, S. T., de Antonio, A., Ferré, X., López, M., Maté, L.: The Software Process: Modelling, Evaluation and Improvement. In: *Handbook of Software Engineering and Knowledge Engineering* (S. K. Chang, ed.), Volume 1, World Scientific Publishing (2001), 193-237.
4. Althoff, K.-D.: Case-Based Reasoning. In: *Handbook of Software Engineering and Knowledge Engineering* (S. K. Chang, ed.), Volume 1, World Scientific Publishing (2001), 549-588.
5. Althoff, K.-D., Pfahl, D.: Making Software Engineering Competence Development Sustained through Systematic Experience Management. In: *Managing Software Engineering Knowledge* (A. Aurum, R. Jeffery, C. Wohlin and M. Handzic, eds.), Berlin: Springer (2003), 269-294.

6. Angkasaputra, N., Pfahl, D.: Making Software Process Simulation Modeling Agile and Pattern-based. In: *Proceedings of the 5th International Workshop on Software Process Simulation Modeling*, ProSim 2004, Edinburgh, Scotland (2004), 222-227.
7. Basili, V., Caldiera, G., Rombach, D.: Experience Factory. In: *Encyclopedia of Software Engineering* (J. Marciniak, ed.), Volume 1, John Wiley & Sons (2001), 511-519.
8. Basili, V., Caldiera, G., Rombach, D., Van Solingen, R.: Goal Question Metric Paradigm. In: *Encyclopedia of Software Engineering* (J. Marciniak, ed.), Volume 1, John Wiley & Sons, (2001), 578-583.
9. Christie, A.M.: Simulation: An Enabling Technology in Software Engineering. In: *CROSSTALK – The Journal of Defense Software Engineering* (1999), 2-7.
10. Curtis, B., Kellner, M. I., Over, J.: Process Modeling. In: *Communications of the ACM* 35 (1992), 9, 75-90.
11. Forrester, J.W.: *Industrial Dynamics*. Productivity Press, Cambridge (1961).
12. Lee, B., Miller, J.: Multi-Project Management in Software Engineering Using Simulation Modeling. In: *Software Quality Journal* 12 (2004), 59-82.
13. Miller, J., Macdonald, F., Ferguson, J.: ASSISTing Management Decisions in the Software Inspection Process. In: *Information Technology and Management* 3 (2002), 67-83.
14. Mehner, T., Messer, T., Paul, P., Paulisch, F., Schless, P., Völker, A.: Siemens Process Assessment and Improvement Approaches: Experiences and Benefits. In: *Proceedings of the 22nd Computer Software and Applications Conference* (COMPSAC), Vienna (1998).
15. Münch, J.: Pattern-based Creation of Software Development Plans. *PhD Theses in Experimental Software Engineering*, Vol. 10 (2001), Fraunhofer IRB, Stuttgart, Germany.
16. Pfahl, D.: An Integrated Approach to Simulation-Based Learning in Support of Strategic and Project Management in Software Organisations. *PhD Theses in Experimental Software Engineering*, Vol. 8 (2001), Fraunhofer IRB, Stuttgart, Germany.
17. Pfahl, D., Lebsanft, K.: Using Simulation to Analyse the Impact of Software Requirement Volatility on Project Performance. In: *Information and Software Technology* 42 (2000), 14, 1001-1008.
18. Pfahl, D., Ruhe, G.: System Dynamics as an Enabling Technology for Learning in Software Organisations. In: *Proceedings of 13th International Conference on Software Engineering and Knowledge Engineering*, SEKE 2001. Skokie: Knowledge Systems Institute (2001), 355-362.
19. Pfahl, D., Ruhe, G.: IMMoS - A Methodology for Integrated Measurement, Modelling, and Simulation. In: *Software Process and Improvement* 7 (2002), 189-210.

20. Pfahl, D., Ruhe, G.: System Dynamics and Goal-Oriented Measurement: A Hybrid Approach. In: *Handbook of Software Engineering and Knowledge Engineering* (S. K. Chang, ed.), Vol. 3, Skokie: Knowledge Systems Institute (to appear).
21. Pfahl, D., Stupperich, M., Krivobokova, T.: PL-SIM: A Simulation Model for Studying Strategic SPI in the Automotive Industry. In: *Proceedings of the 5th International Workshop on Software Process Simulation Modeling*, ProSim 2004, Edinburgh, Scotland (2004), 149-158.
22. Kellner M. I., Madachy, R. J., Raffo, D. M.: Software process simulation modeling: Why? What? How?. In: *Journal of Systems and Software* 46 (1999), 91-105.
23. Rus, I., Collofello, J. S.: A Decision Support System for Software Reliability Engineering Strategy Selection. In: *Proceedings of the 23rd Annual International Computer Software and Applications*, COMPSAC 99, Scottsdale, AZ, October 1999, 376-381.
24. Ruhe, G., Ngo-The, A.: Hybrid Intelligence in Software Release Planning. In: *International Journal on Intelligent Hybrid Systems*, Vol 1 (2004), 99-110.
25. Ruhe, G.: Software Engineering Decision Support - A New Paradigm for Learning Software Organizations. In: *Advances in Learning Software Organization.* Lecture Notes In Computer Science Vol. 2640, Springer (2003), 104-115.
26. van Solingen, R., Berghout, E.: *The Goal/Question/Metric method: A practical guide for quality improvement of software development.* McGraw-Hill Publishers (1999).
27. Schneider, K.: Experience-Based Process Improvement. In: *Proceedings of 7th European Conference on Software Quality*. Helsinki, Finland (2002).
28. Shim, J. P., et al.: Past, present, and future of decision support technology. In: *Decision Support Systems* 33 (2002), 111-126.
29. *Ventana Simulation Environment (Vensim®) - Reference Manual*, Version 5.0b, Ventana Systems, Inc. (2002).
30. Völker, A.: Software Process Assessments at Siemens as a Basis for Process Improvement in Industry. In: *Proceedings of the ISCN*, Dublin, Ireland (1994).
31. Waeselinck, H., Pfahl, D.: System Dynamics Applied to the Modelling of Software Projects. In: *Software Concepts and Tools* 15 (1994), 4, 162-176.

Chapter 4

HIGH LEVEL SOFTWARE PROJECT MODELING WITH SYSTEM DYNAMICS

Márcio De Oliveira Barros[1,2], Cláudia Maria Lima Werner[2], Guilherme Horta Travassos[2]

[1] *DIA / UNIRIO – Applied Informatics Department*
Av.Pasteur 458, Térreo – Rio de Janeiro – RJ – CEP:22290-240
E-mail: marcio.barros@uniriotec.br

[2] *COPPE / UFRJ – System Engineering and Computer Science Department*
PO Box: 68511 - CEP 21942-970 - Rio de Janeiro – RJ,
E-mail: werner@cos.ufrj.br, ght@cos.ufrj.br

System dynamics based software project models are becoming increasingly more complex due to continuing research on the field. Current models usually convey several hundreds or even thousands of equations. This amount of formal knowledge is hard to maintain and adapt for circumstances other than those for which it was originally developed. So, it is common that only experts build and adapt system dynamics project models. In this chapter, we present an approach based on high level modeling for software projects, which separates the description of a particular project from the knowledge embedded in a software project model.

1. The Need for High Level Modeling

Software project models based on system dynamics[1, 2, 3, 4, 5] convey representations for agents and artifacts that participate in a software development effort. Traditionally, mathematical equations describing system dynamics constructors (stocks, rates, processes, and flows) are used to represent such models[6]. Although these constructors are flexible, they are also fine-grained, and, as system dynamics models grow, they become hard to understand. Software project elements, such as developers, activities, and artifacts, are not easily identified in a maze of constructors. Their representation is usually spread across several equations, thus forcing modelers to analyze the whole model to determine the precise group of equations that describe the behavior of an element and its relationships to others.

Each software project model embeds generic and reusable knowledge about the software project domain. However, there is no clear separation between domain knowledge and the characteristics related to a particular project in traditional system dynamics models: both are distributed across the model equations. This characteristic limits our ability to reuse domain knowledge, leading to some limitations in the traditional modeling approach. They include (1) inhibiting the creation of large reusable models, since the relationships among elements in large models are difficult to understand and observe from equations; and (2) reducing economy of scale, which is obtained when each new model developed for a specific domain requires less effort than the previous models developed for the same domain. Since model reuse is inhibited, modelers spend some effort reacquiring potentially reusable domain knowledge for each new model developed for a domain.

System dynamics models also tend to describe uniformly all elements pertaining to the same category in the modeling domain by using average values to describe such elements. For instance, it is very common to assume that all experienced developers have the same productivity and generate the same amount of errors per thousand lines of code[1]. Generally, this simplification is due to system dynamics' inherent inability to describe element properties, since these should be independent model variables, requiring too many equations to be specified.

Finally, system dynamics models usually blend known facts about the elements within a model with uncertain assumptions about their behavior. In decision support models, known facts may represent the expected model behavior if no action is taken, while each assumption may represent how the model reacts when a particular decision (and the actions triggered by this decision) is taken. In software project models, for instance, an assumption may represent how coding activities are affected when inspections are accomplished during the development process. By partitioning a model into known facts and assumptions, a modeler can analyze the model behavior in alternative situations, according to distinct combinations of such assumptions.

While we need models that are simple to understand, we also want models that can represent the details of their interacting elements. However, as we add details to system dynamics models they usually require more equations, thus leading to difficulties in understanding and maintenance. So, we perceive the need for techniques to enhance the development of such complex models. If domain knowledge can be clearly separated from the particular problem information, every model developed for that domain can be reused. By reusing domain information created and organized by previous models, the cost of developing new models within a domain can be reduced. If conditional or uncertain behavior can be separated from expected behavior, we can analyze a model according to different sets of assumptions. Moreover, if a modeler builds a scenario model describing an uncertain behavior associated to a domain (instead of a specific model), the model can be reused in the analysis of several distinct models developed for the domain.

In this chapter, we present a software project modeling approach that handles model complexity by raising the abstraction level of modeling constructors. Several models are developed applying the proposed approach. Many of these models are built by software project modeling experts: they use traditional system dynamics constructors to describe high-level concepts of the problem domain (such as activities, developers, and artifacts) and embed their knowledge about such elements. Project managers use the former concepts to build models to represent their specific projects. Since not every project manager is a modeling expert (or can usually afford the time required to build project models), their models are built upon the high level constructors developed by the experts. Equations (elements of the solution domain and embedded in the high level concepts by the experts) are required for model simulation and analysis, but they are not best suited for model description, since they represent concepts far from the problem domain. So, the model built by the project manager is expressed in a language that is closer to the manager perspective, being further translated to mathematical representation.

Our approach is composed of a system dynamics metamodel, scenario models, and a translation process, which is guided by a modeling process. The metamodel is the language that allows the description of categories of elements that collaborate within a problem domain and the relationships among them. Scenario models represent extensions to the domain model, which record knowledge about theories, events, practices, and strategies that cannot be presumed true for every project, but which can hold in specific situations. The translation process compiles the metamodel and scenario model representation into system dynamics constructors, which can be further used for simulation and model evaluation. Model behavior is expressed using extended system dynamics constructors, which are separately described for each element category within the domain.

This chapter is organized in 7 sections. The first one comprises this introduction to the need for high-level modeling. The next section presents some concepts that will be used throughout the chapter. In section 3, we present the modeling process that drives the application of the system dynamics metamodel. In section 4, we present an application of the modeling process. In section 5, we present the simulation process that uses the artifacts produced by the modeling process. In section 6, we present an application of the simulation process. Finally, in section 7 we discuss the application of the system dynamics metamodel and scenario models, presenting our final considerations and conclusions.

2. Definitions

Before presenting the system dynamics metamodel and scenario models, we have to define some terms that will be used throughout the following sections. Some of this terminology comes from a subset of the object-oriented software development theory[7]. However, such terms need to be adapted to the system dynamics context.

A **class** represents a set of elements that can be described by the same properties and exhibit similar behavior. For instance, a class may describe the whole group of software developers, while each particular developer is an **instance** of the class.

A class defines the properties that describe its instances. A **property** represents relevant information about a class. Each instance assumes an independent value for each property defined in the class, according to the characteristics of the element represented by the instance.

A class also defines the behavior of its instances. The **behavior** of a class is a mathematical formulation of its responses to changes in other instances or in the environment. Such behavior can depend on class properties, allowing distinct class instances to react differently based on their particular characteristics or state. System dynamics constructors describe class behavior.

Class instances can have relationships to other class instances. A **relationship** represents a structural connection between two or more class instances. Such relationships can occur among instances of different classes or instances of the same class. The latter is also called an **auto-relationship**. A **role** represents the part that an instance undertakes in a relationship. It denotes the responsibilities and expected instance behavior.

A **domain model** contains classes for the elements that participate in a problem domain, describing their properties, behavior, and the potential relationships among their instances. The domain model does not describe a model for a specific problem, but a knowledge area where modeling can be applied. It is a generic domain description, which should be specialized in every attempt to model a problem within the domain.

A **scenario model** extends a domain model by providing new behavior and properties for one or more classes defined within the domain. A scenario model is composed of connections and constraints.

A **connection** is an association between a scenario model and a class within a domain model. The connection allows the scenario to be enacted upon instances of the class in a specific model for the domain.

The enactment of a scenario model upon an instance is an **activation**. An activation implies that the equations defined within the scenario model are included in the behavior of that particular instance of a class, thus modifying its original behavior.

A **constraint** is an activation restriction that an instance and its associated instances have to attend in order to be connected to the scenario. The restriction states that the instance itself or its associated instances must have a particular scenario connection enacted upon them. This restriction allows a scenario to use or modify the behavior provided by a second scenario, preventing a modeler from enacting its connections upon an instance that is not connected to the second scenario.

3. The Modeling Process

We propose a four-step modeling process. First, an expert in a given domain (such as the software development process domain) builds a domain model. This model cannot be simulated, since it does not specify how many instances of each class exist in the model, nor does it specify any value for their properties. This step is called **domain modeling**. This is the most expensive step in the modeling process, since properties and behavior equations must be specified for each domain class. However, an investment in developing a high quality domain model may payoff in the future, when knowledge expressed in this model can be reused, reducing development time for modeling specific software projects.

The creation of a model based on a domain model is the second step in the modeling process. At this time, a modeler specifies how many instances of each class defined for the domain exist in the model of interest. The modeler also specifies the values for instance properties and describes how these instances are related to each other, according to the domain model relationships among classes. This step is called **model instantiation**. The resulting model only conveys information about its instances: it does not present any system dynamics constructor. So, the high-level representation helps model development and understanding.

At the third step, the model that indicates the relevant instances of the domain model classes (built in step 2) is translated to system dynamics equations. This step is called **model compilation**. The resulting model uses only standard system dynamics constructors, while the input model is described in the high level representation. While the high level model is supposed to be easier to develop, it is the constructor-based representation that allows simulation and behavior analysis.

The first three steps of the modeling process resemble a cascading pattern, where each activity is executed after its preceding activity's conclusion. The fourth activity does not take part in this linear execution pattern. Instead, the **scenario model development** activity is executed when there is an opportunity to build a scenario model. In the software development context, this opportunity arises when an experienced manager decides to represent some knowledge about a software project in a form that can be reused and applied by other managers while analyzing their specific models. A scenario model can only be developed after the conclusion of the domain modeling activity, when a domain model is available. The scenario modeler studies the domain model and any other available scenario models to determine which equations should be affected by the new scenario model to provide the desired behavior. The scenario model is written and stored in a corporate repository, from where it can be retrieved later.

We expect a particular problem within the domain (e.g., a particular software project) to be easier to model applying this modeling process than using pure system dynamics constructors, since modelers will use domain concepts described by the domain-specific language (the domain model) to build their models. Similar approaches have been used in other areas, such as domain model description languages[8].

4. An Application of the Modeling Process

In this section, we show an application of the modeling process presented in section 3 to create a model for a small software project. The project is described in Figure 1, which shows developers participating in software development activities. The lines connecting activities and developers represent the relationships among these elements.

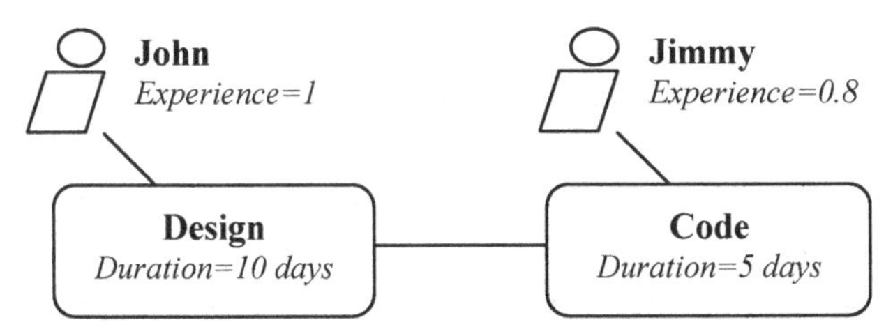

Figure 1. Small software process example

In Figure 1, two developers, namely *John* and *Jimmy*, design and code a software module. Developers have experience, shown as a rate in the unit interval, which influences their work quality. Activities have an estimated duration. This specifies how many days of work it is expected to take an average developer to complete these activities.

The following sections discuss how a domain model can represent such a problem domain, how a specific model represents the project itself, and how this model can be compiled into the traditional system dynamics representation for simulation. Although scenario modeling was described as part of the modeling process (fourth activity), we will leave the discussion regarding scenario models to the sections that describe the simulation process.

4.1. Domain Modeling

To exemplify the development of a domain model for the software project domain, consider that the relevant classes within the domain are just activities and developers. Table 1 presents a simplified model for the software project domain using the concepts defined in section 2 and a language to represent such concepts.

The **MODEL** keyword introduces the domain model, namely *SoftwareProject*. It contains two classes, each one declared by using the **CLASS** keyword. Classes are declared within the domain model context, delimited by angled brackets. Each class contains its own context, where properties and behavior are declared.

The **PROPERTY** keyword specifies a property for a class. Default property values are defined next to the property name. For instance, the domain model defines the *Experience* property for the *Developer* class. This property describes a developer's ability to accomplish activities. When developing a particular model for the domain (step two), the modeler must determine how many developers are needed and specify each developer's experience level. Otherwise, for those instances in which experience level is not defined, the default value will be assumed. If precision is not required, an average value can be defined as the default value for a property, and instances where the property is not specified will assume this value. The *Activity* class conveys a single property, *Duration*, which specifies how long it takes for an average developer to do the activity.

Table 1. A simple domain model for the software project knowledge area

```
MODEL SoftwareProject
{
  CLASS Developer
  {
    PROPERTY Experience 1;
    PROC Productivity Experience;
    PROC ErrorGenRate 1;
  };

  CLASS Activity
  {
    PROPERTY Duration 0;

    PROC Prod Team.Productivity;
    STOCK TimeToConclude duration;
    RATE (TimeToConclude) Work if(DependOk, -Prod * Min
        (TimeToConclude/Prod, 1)/DT, 0);
    PROC DependOk GROUPSUM (Precedence, TimeToConclude) < 0.001;

    STOCK Executing 0;
    RATE (Executing) RTExecuting if (AND(Executing < 0.001,
        DependOk), 1, 0);

    PROC InErrors GROUPSUM(Precedence, Errors);
    RATE (Errors) ErrorsTransmit if (RTExecuting>0.001, InErrors
        / DT, 0);

    STOCK Errors 0;
    PROC ErrorsInTask 10 * Team.ErrorGenRate;
    RATE (Errors) ErrorsCommited -ErrorsInTask * (Work /
        Duration);
  };

  RELATION Team Activity, Developer (MyActivities);
  MULTIRELATION Precedence Activity, Activity (NextActivities);
};
```

A relationship between two or more instances allows one instance behavior equation to access and even modify another instance behavior (for instance, a rate equation in a class behavior can affect stocks defined in other classes). During model compilation, model equations referencing such relationships are adjusted to the configuration of instances and connections presented in the model. The domain model allows two types of relationships among classes:

- Single relationships (also known as 1:1 associations): one instance of a source class is associated to a single instance of a target class;

- Multi-relationships (also known as 1:N associations): one instance of a source class is associated to several instances of a target class.

Consider that, after observing the model behavior, a manager decides to try different staff allocations to project activities. Profound changes would have to be made to the model written without using the domain model for each staff allocation analysis, since the relationships between developers and activities are hard-coded within system dynamics equations. By using the proposed modeling process and notation, the modeler would just have to change these relationships, which are clearly stated for each model instance.

The **RELATION** keyword represents a single relationship, such as *Team*, which denotes that an activity is executed by a single developer. The **MULTIRELATION** keyword represents a multiple relationship, such as *Precedence*, which represents the activities that must be concluded before another activity can start.

Relationships are unidirectional by default, that is, only the source instance has access to the target behavior. For instance, according to the *Team* relationship, only the *Activity* class instance has access to the information about its developers. Relationships can be set to bi-directional by specifying a role for the target instance. The target instance can manipulate the behavior of its source instance through this role name. The role is specified within parenthesis next to the target class name. The *Team* relationship is single and unidirectional. The *Precedence* relationship is a multiple, bi-directional auto-relationship, since it links instances of the same class. The *NextActivities* role is specified for the target class, as required by bi-directional relationships.

Behavior equations at domain model level are distributed among several classes, each class containing its specific behavior. The *Developer* class has very simple behavior. It only defines two processes: one to store its *Experience* property value, and a second to state the developer's error generation rate as a unitary value. The first process allows other instances to consult the property value, since the property itself can only be accessed by its instance.

The *Activity* class contains most of the behavioral equations presented in the domain model. The *TimeToConclude* stock describes the time required for an activity to be accomplished, being depleted by the *Work* rate as the simulation advances. Observe that the stock name, presented in the rate within parenthesis after the **RATE** keyword, associates the rate to the stock. Rates are always supposed to raise their associated stock level in the domain model. Rate equations must generate negative numbers (as happens in the *Work* rate) to allow stock depletion.

For the purpose of this example, an activity can only be executed when all preceding activities have been concluded. So, the *Work* rate depends on the *DependOk* process, which determines if the preceding activities of an activity are already concluded, based on the *Precedence* relationship. This process uses the **GROUPSUM** operator, which sums the values of a selected property for every instance associated to the current instance through a specific relationship. In the *DependOk* process, the operator sums the *TimeToConclude* stock level for every activity that must be executed before the current one. The *DependOk* process verifies if the operation result is near to zero, determining whether the activities have already been accomplished.

The next two equations (*Executing* and *RTExecuting*) are used to create a variable that contains zero most of the time, but raises to one in the simulation step that marks an activity start. This variable is used by the *ErrorsTransmit* rate, which triggers the transmission of errors produced in preceding activities to the current one, thus raising the *Errors* stock. In the example, we assume that all errors that exist in the artifacts produced by preceding activities will be reproduced in the artifact produced by the current activity. The *Errors* stock represents the number of errors produced by an activity. It starts with zero errors, being affected by the *ErrorsTransmit* and *ErrosCommited* rates. The second rate adds new errors to the artifact produced by the activity, according to the assigned developer error generation rate. For the sake of simplicity, we assume that an average developer generates 10 errors per task.

4.2. Model Instantiation

Table 2 presents a model for the software project presented in Figure 1. Note that the model is not built upon equations or a traditional system dynamics constructor. Instead, it uses the classes declared in the domain model, creating instances for these classes, and specifying their property values and relationships to other instances.

The **DEFINE** keyword introduces the project model, followed by the model name (*MyProcess*) and by the domain model to which it is related (*SoftwareProject*). Class instances are represented within the model context, delimited by angled brackets.

Table 2. Specific model for the software project domain

```
DEFINE MyProcess SoftwareProject
{
   John = NEW Developer
      SET Experience = 1;

   Jimmy = NEW Developer
      SET Experience = 0.8;

   Designing = NEW Activity
      SET duration = 10;
      LINK Team John;

   Coding = NEW Activity
      SET duration = 5;
      LINK Team Jimmy;
      LINK Precedence Designing;
};
```

The developers (*John* and *Jimmy*) are the first instances presented within the model. The **NEW** keyword creates a class instance identified by the name presented after the keyword. The newly created instance is associated to the identifier on the left side of the equal operator. Next, class instances are presented for the artifacts and the activities within the software project.

The **SET** keyword specifies a property value for a specific instance. *John's* experience is supposed to be 1, while *Jimmy's* experience is supposed to be 0.8. This feature allows modelers to precisely account for the relevant differences between instances of the same class, which requires several equations in the traditional system dynamics models. If property values are not specified for an instance, the default value defined in the domain model is used.

The model also presents the occurrences of the *Precedence* and *Team* relationships. Only the activities specify relationships, since they are always referenced as source classes. The **LINK** keyword determines which class instances are associated in each relationship. For instance, the *Coding* activity is dependent on the *Designing* activity and is developed by *Jimmy*.

4.3. Model Compilation to System Dynamics Constructors

The techniques presented in the previous sections are an aid for constructing larger and detailed models, but they would be rendered useless if these models could not be simulated. Class-based representation must be translated to system dynamics constructors to provide simulation capability, which can be analyzed in a conventional simulator. We have developed a compiler, named Hector[9], which translates the high-level description presented in the preceding sections to traditional system dynamics constructors. Table 3 presents an extract of the compiled version of the model presented in section 4.2. Lines are numbered to support further discussion in the remainder of this section.

The compiled model only conveys system dynamics constructors, which are represented using the ILLIUM tool modeling language[10]. This language allows the definition of stocks, rates, and processes. Every constructor has a unique name, used to identify it in the model, and an expression, which is evaluated in every simulation step. Rates are also associated to two stocks, which represent the origin and the target of its flow. Universal providers[I], represented by the SOURCE keyword, or universal sinkers, represented by the SINK keyword, can replace such stocks.

To avoid confusion we will refer to the class-based representation as model, while the system dynamics-constructors based version will be

---

[I] Universal providers and sinkers represent the limits of a model, usually depicted as clouds in stock-and-flow diagrams.

referred to as compiled model. The extract of the compiled model presented in Table 3 conveys two blocks, describing one instance of the *Developer* class and one instance of the *Activity* class.

Table 3. Extract from the traditional system dynamics model generated from the model presented in section 4.2

```
01 # Code for object "Jimmy"
02 PROC Jimmy_Experience 0.800000;
03 PROC Jimmy_Productivity Jimmy_Experience;
04 PROC Jimmy_ErrorGenRate 1.000000;

05 # Code for object "Coding"
06 PROC Coding_Duration 5.000000;
07 PROC Coding_Prod Jimmy_Productivity;
08 STOCK Coding_TimeToConclude Coding_duration;
09 RATE (SOURCE, Coding_TimeToConclude) Coding_Work IF
       (Coding_DependOk, -Coding_Prod * MIN (Coding_TimeToConclude /
       Coding_Prod, 1.000000) / DT, 0.000000);
10 PROC Coding_DependOk (Designing_TimeToConclude) < 0.001000;
11 STOCK Coding_Executing 0.000000;
12 RATE (SOURCE, Coding_Executing) Coding_RTExecuting IF (AND
       (Coding_Executing < 0.001000, Coding_DependOk), 1.000000, 0.000000);
13 PROC Coding_InErrors (Designing_Errors);
14 RATE (SOURCE, Coding_Errors) Coding_ErrorsTransmit IF
       (Coding_RTExecuting > 0.001000, Coding_InErrors / DT, 0.000000);
15 STOCK Coding_Errors 0.000000;
16 PROC Coding_ErrorsInTask 10.000000 * Jimmy_ErrorGenRate;
17 RATE (SOURCE, Coding_Errors) Coding_ErrorsCommited -
       Coding_ErrorsInTask * (Coding_Work / Coding_Duration);
```

Consider the equations generated for the *Jimmy* instance of the *Developer* class. Line 01 only contains a compiler-generated comment. The equations from line 02 to 04 convey the declaration of a property and two behavior constructors. The first equation declares the *Experience* property for the *Jimmy* instance. Since every class instance may have different property values, every property must be represented by an independent equation. Several equations are required to represent the whole set of instances, capturing their particular properties. This leads to larger models that would be error-prone if modelers decide to write the equations directly. By using the domain model and the system dynamics metamodel approach, modelers are allowed to define different values for each instance properties and instance behavior is automatically adjusted to these values during the model compilation step.

Properties are declared as processes in the compiled model. They are initialized with the value specified for them in the model or by their default value, as stated in the domain model. Notice that the name of the process representing the property in the compiled model is composed of the instance name followed by the property name. Both names, separated by an underlining sign, make up a unique identifier, which serves as a name for the process within the compiled model. This allows the declaration of several instances with distinct property values, since each instance is required to have a unique name in the instance-based model.

The second and third equations represent the behavior equations defined in the *Developer* class, which are specialized for the *Jimmy* instance. References to properties in the behavior equations are linked to the processes that represent such properties in the current instance. The instance name is also used as a prefix to the behavior constructor name in the compiled model. Behavior descriptions are repeated for every instance in the compiled model.

The model generated for the *Coding* instance of the *Activity* class is more interesting. Line 05 contains a compiler-generated comment. The first equation generated for this instance declares the *Duration* property (line 06). As in the *Jimmy* instance's generated code, a process is used to represent the property and its value.

The next equation (line 07) presents a behavior equation that is parameterized by a relationship. It represents the *Prod* process, which is defined in the domain model as the productivity of the developer associated to the current activity through the *Team* relationship. Since *Jimmy* plays this role in the current model (see the **LINK** keyword in the activity declaration), its productivity equation is used in the *Prod* process. The compiler resolves relationship dependences by exchanging the relationship name for the instance names that play the role described in the relationship. In line 07, the compiler exchanges the relationship name as declared in the metamodel (*Team*) for the name of the developer that is associated with the activity in the instance-based model (*Jimmy*).

The following equation (line 08) conveys the *TimeToConclude* stock for the *Coding* activity, which represents the number of days required to conclude this activity. This stock is depleted by the *Work* rate (line 09). This rate checks the *DependOk* process (line 10) to determine whether its precedent activities are concluded. The process uses the GROUPSUM operator, which is compiled to a list of arithmetic sums whose operands are the instances participating in the relationship selected for the operator. The *DependOk* behavior description within the Activity class uses the Precedence relationship. In the *Coding* activity, which is preceded by a single activity (*Designing*), the GROUPSUM operator is compiled as a reference to a behavior equation of the *Designing* instance. In the *Designing* activity the GROUPSUM operator is compiled to zero, since there is no precedent activity (therefore, resulting in no operand for arithmetic sums).

The *DependOk* behavior within the Coding instance uses a stock declared by other instance (in this case, the *Designing* activity). This is accomplished through model relationships, which allow an instance to consult or modify other instances' behavior. The compiling process perceives such access to externally defined behavior through the relationship name. So, it puts the name of the instance being manipulated in front of the name of the accessed behavior.

Further equations represent activity execution time (lines 11 and 12), error transmission from precedent activities (lines 13 and 14), and errors generated while the activity is executed (lines 15 to 17). These equations were compiled using the same strategies as described earlier in this section.

5. The Simulation Process

The artifacts developed by an execution of the modeling process are used during the simulation process. This process is executed to support decision making for a specific software project. It allows an analyst, who is usually a project manager, to observe the behavior of a software project according to specific situations.

The simulation process starts after the compilation of the specific model that describes a software project (third activity of the modeling process). The compiled model is executed in a simulator and the analyst can observe its behavior, which is described in reports and graphical plots. This model acts as a baseline, containing known facts about a software project without any uncertain assumptions. The baseline model behavior shows what is expected to happen during project development if the project is not affected by uncertain events (for instance, sick developers, unmotivated team, problems in the working environment, higher error generation rates, higher requirements volatility, lack of support from senior management, and so on).

However, no software project can be isolated from every kind of uncertain events, and the analyst must study how the project reacts to the occurrence of such events. Scenario models represent uncertain events, so the analyst can retrieve these models from the corporate repository. If models for the specific uncertain events under interest are not available, the analyst can either build such models (by applying the fourth activity of the modeling process) or abort the analysis of these events, focusing on events that were previously documented as scenario models.

After retrieving or building scenario models, the analyst performs an iterative process where these models are integrated into the specific model developed for the software project of interest. Each combined model is simulated to evaluate how the assumptions documented in the scenario models affect the baseline model behavior. The scenario model integration process does not require manual changes to the software project model equations, thus reducing the chances of errors being introduced in the model. Also, since scenarios are described as separate models, they tend to be easier to reuse than groups of equations manually extracted from a model.

In decision support models, the results provided by the simulation executed before scenario model integration represent the expected model behavior if the underlying process is unaffected by external decisions. The simulations done after scenario integration show the expected model behavior when each particular decision (represented by a scenario model) is taken.

6. An Application of the Simulation Process

In this section, we show an application of the simulation process based on the models developed in section 4. The small software project introduced in that section will be further used to support our discussions.

6.1. Scenario Models

Scenario models extend domain models by providing new behavior and characterization for one or more domain classes. A scenario model is composed of connections and constraints. A *connection* associates the scenario to a domain class, so that the scenario can be enacted on instances of the class in a specific model for the domain. A *constraint* declares restrictions that the connected instances and its associated instances have to meet in order to be connected to the scenario.

Table 4 presents a scenario model for the software project domain. This scenario describes the behavior of an overworking developer, that is, a developer that works more than eight hours per day. Its equations were adapted from Abdel-Hamid and Madnick[1], according to the heuristic first presented by DeMarco[11] that states that overworking developers work more (they present higher productivity), but not better (they present higher error generation rates). The *Overworking* scenario has a single connection and no constraints.

A connection extends the behavior and characterization of a domain model class by adding new properties and new behavior equations to the class. These properties and behavior equations are intended to be relevant to the model only if the assumption described by the scenario is relevant to the modeler. Thus, the number of hours that each developer works per day must only be specified if the analyst wants to measure the impact of overworking upon the project model. If this scenario is not relevant to the analyst, the scenario model is not considered, and its properties and behavior equations are not taken into account during the model simulation. The *TheDeveloper* connection allows the *Overworking* scenario to be connected to the *Developer* class in the *SoftwareProject* domain. This connection declares a single property (*WorkHours*), which indicates how many hours a developer works per day, and three behavior equations, represented by the processes and stocks in the scenario model.

Table 4. A scenario model for the software project domain

```
SCENARIO Overworking ProjectModel
{
  CONNECTION TheDeveloper Developer
  {
    PROPERTY WorkHours 8;    # 8 to 12 hours

    STOCK DailyWorkHours WorkHours;
    PROC  WHModifier 1 + (DailyWorkHours - 8) / (12 - 8);

    PROC  SEModifier LOOKUP (SchErrorsTable, WHModifier-1, 0, 1);
    TABLE SchErrorsTable 0.9, 0.94, 1, 1.05, 1.14, 1.24, 1.36, 1.5;

    AFFECT Productivity Productivity * WHModifier;
    AFFECT ErrorGenRate ErrorGenRate * SEModifier;
  };
};
```

Connections also declare behavior redefinition clauses, which allow the scenario to change the equations of rates and processes defined for its associated class in the domain model. The *Overworking* scenario has two behavior redefinition clauses, represented by the *AFFECT* keyword. The first redefinition clause indicates the raise in productivity due to overworking, while the second redefinition clause indicates the expected effects of overworking upon developers' error generation rate. The original equations that describe developer's productivity and error generation rate are overridden by scenario definitions. For instance, the *Overworking* scenario redefines the *Productivity* equation in the developer class by multiplying its original value by a factor that depends on the number of hours that the developer works per day.

6.2. Scenario Activation

A scenario is not a self-contained model. It is a complementary model that adjusts the behavior of previously developed models. It can be activated upon specific models developed for the same domain to which the scenario was created. When a scenario is activated upon a model, its connections must be enacted upon class instances declared within the specific model. The effects of enacting a connection upon an instance are similar to declaring the properties and behavior equations that are defined in the connection directly in the domain model class. However, if such properties and behavior equations were declared in the domain model, they would apply for every instance of the class in every specific model developed for the domain. Scenario connections can be enacted upon specific class instances, modifying the behavior of those particular instances. The remaining instances in the model present only their original class behavior and properties, without the effects provided by the scenario. Table 5 shows an *Overworking* scenario activation upon developer's instances in the model shown in Table 2.

Table 5. Scenario model activation upon a model for the software project domain

```
DEFINE MyProcess ProcessModel
{
  John = NEW Developer
    SET Experience = 1;
    SET WorkHours = 12;

  Jimmy = NEW Developer
    SET Experience = 0.8;
    SET WorkHours = 8;
  ...

  ACTIVATE Overworking
    CONNECT TheDeveloper Jimmy;

  ACTIVATE Overworking
    CONNECT TheDeveloper John;
};
```

Properties defined by a scenario connection are added to the list of properties that describe the class instance upon which the connection was enacted. As in the domain model, these properties have a default value, which can be redefined by particular instances in a model. Notice that the instance that represents the developer named *John* redefines the value of the *WorkHours* property in Table 5. Connection equations assume the new property value for the instance, adjusting scenario behavior for this value. If no scenario connections were enacted upon the instance, the initialization of the *WorkHours* property would result in an error, since the property was not defined for the class in the domain model.

The *Overworking* scenario and the *SoftwareProject* domain model illustrate the major advantage of using scenario models, that is, separating the hypothesis concerning a theory (represented in the scenario model) from the known facts represented in a specific model. Consider that a modeler wants to evaluate the number of errors over time in the example project in two distinct situations: without overworking and considering the effects of overworking upon the project behavior. To evaluate the first situation, the modeler uses the model for the *SoftwareProject* domain shown in Table 2. The number of errors in the project over time in this model is shown in the left-hand graph in Figure 2. Next, to evaluate the overworking behavior, the modeler activates the *Overworking* scenario upon the developers, as presented in Table 4. This model shows the resulting behavior in the right-hand graph in Figure 2 (all other variables were unchanged).

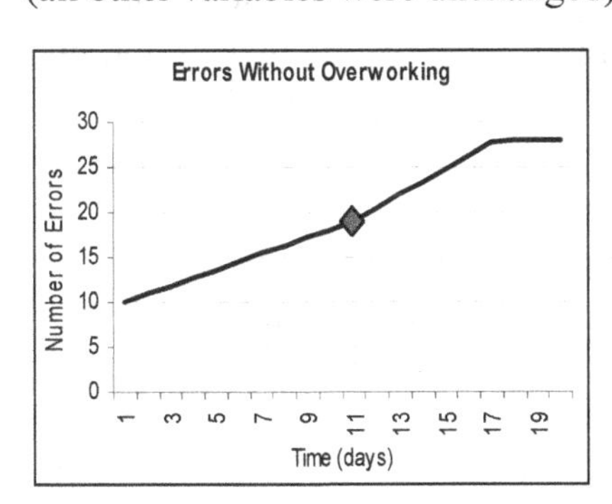

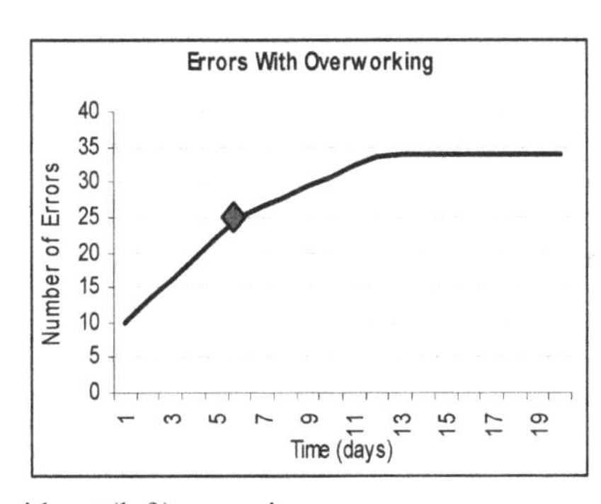

Figure 2. Model behavior with (right) and without (left) scenarios

The diamond indicates the number of errors by the end of the *Designing* activity, which was executed by developer *John* working twelve hours per day. Notice that the activity is concluded faster (day 6 versus day 12) when the *Overworking* scenario is taken into account, due to a higher productivity rate. However, the number of errors also increases in the analysis with the *Overworking* scenario: 25 errors by the end of the *Designing* activity against only 19 errors without overworking. Moreover, since error detection and correction are not included in the example project, it is concluded sooner, though with lower quality, when overworking is considered (12 days versus 17 days).

Thus, scenarios allow a modeler to perform behavior analysis upon a model without direct intervention in its equations or control parameters. Scenarios act as "plug-&-simulate" extensions to a domain model, providing different behavior for classes that can be plugged and analyzed, according to the modeler's needs. These analyses can be rather difficult using the textual model representation for scenarios and models presented in the preceding examples. In this context, graphical tools to support the system dynamics metamodel, scenario creation and integration can be built to allow a user to graphically build a model and activate scenarios upon it. Currently, the Hector metamodel compiler manages scenario activations when transforming a specific model to traditional system dynamics equations[9].

6.3. Constrained Scenarios

Scenarios may depend on other scenarios to represent their behavior. Suppose we have a scenario that represents developers' exhaustion due to overworking during the development of a project. The scenario presented in Table 6 represents this effect. Its equations were adapted from Abdel-Hamid and Madnick[1].

Whereas the *Overworking* scenario presented in Section 6.1 focused on the error generation behavior occurring when developers work more than their regular working hours per day, the *Exhaustion* scenario (Table 6) states that developers become tired as they overwork. If the overworking period lasts too long, developers are so exhausted that they refuse to overwork during a "resting" period. While resting, developers work only eight hours per day, despite pressure to overwork.

Within the *Exhaustion* scenario, the *Resting* stock remains zero as long as developers are not exhausted enough to start a resting period. If this stock assumes a non-zero value, it forces the *DailyWorkHours* stock that was presented in the *Overworking* scenario to represent a workday of eight hours. Thus, scenarios that do not present behavior redefinition (*AFFECT*) clauses can affect the behavior of their associated classes by adding rates to stocks previously defined for the classes.

Table 6. A constrained scenario model for the software project domain

```
SCENARIO Exhaustion ProjectModel
{
  CONNECTION TheDeveloper Developer
  {
    STOCK Exhaustion 0;
    PROC MaxExhaustion 50;
    PROC IsResting OR(Resting=1, Groupsum(MyActivities, Work)=0)
    PROC ExhaustionPass Max(-Exhaustion/DT, -MaxExhaustion/20.0);
    RATE (Exhaustion) ExRT if(IsResting, ExhaustionPass,
         EXModifier);

    PROC EXModifier LOOKUP (ExaustionTable, DedicationFactor, 0,
         1.5);
    PROC DedicationFactor 1 - (1 - Dedication) / 0.4;
    PROC Dedication 0.6 + (WHModifier - 1) * (1.2 - 0.6);
    TABLE ExaustionTable 0.0, 0.2, 0.4, 0.6, 0.8, 1.15, 1.3, 1.6,
         1.9, 2.2, 2.5;

    STOCK Resting 0;
    RATE (Resting) RestingRate1 IF (InitResting, 1 / DT, 0);
    RATE (Resting) RestingRate2 IF (QuitResting, -1 / DT, 0);
    RATE (DailyWorkHours) DWHRate IF (Resting = 1, (8 -
         DailyWorkHours) / DT, 0);

    PROC InitResting AND(Resting = 0, Exhaustion > MaxExhaustion);
    PROC QuitResting AND(Resting = 1, Exhaustion < 0.1);
  };

  CONSTRAINT TheDeveloper, Overworking.TheDeveloper;
};
```

However, since the class has not defined the *DailyWorkHours* stock by itself (it was declared by a scenario), the *Exhaustion* scenario can only be activated upon instances of the *Developer* class over which the *Overworking* scenario was previously activated (otherwise, the referenced stock would not be available).

The constraint in the scenario warrants the establishment of this connection. It states that instances affected by the *TheDeveloper* connection in the *Exhaustion* scenario must also be affected by the

TheDeveloper connection of the *Overworking* scenario. If the last connection is not enacted upon the instance, the metamodel compiler issues an error and does not generate the compiled model.

Constraints are not restricted to class instances upon which a scenario connection is enacted. Other instances, linked to the connected instances by class relationships, can also be affected by constraints. To allow associated instances evaluation by a constraint, a dot operator and a relationship identifier should follow the *TheDeveloper* connection on the left-hand side of the comma that divides the constraint declaration. All class instances associated to the connected instance through this relationship should comply with the scenario connection presented by the right-hand side of the comma.

Scenarios are supposed to be small: they should concentrate on the behavior equations that describe a particular problem or opportunity. The power of scenarios is their integration with specific models, and constraints play an important role by stating and verifying the dependencies among scenarios.

6.4. Scenario Activation Ordering

Scenario activation ordering is relevant since several scenarios can redefine the same equation for a class instance in a specific model, and due to operator precedence rules within an equation. If several scenario connections are enacted upon the same class instance, their behavior redefinition clauses affect the original domain class equations according to the scenario activation order.

Consider the hypothetical scenarios presented in Table 7. The first scenario represents a reduction in the error generation rate due, for instance, to learning a new development technique with which a lower number of errors are generated during project development. The scenario indicates that the error generation rate is reduced by a constant factor. The second scenario represents rises in the error generation rates due to, for instance, schedule pressure and a close project conclusion date. The scenario indicates that the error generation rate grows by a multiplying factor. Both scenarios affect a developer's *ErrorGenRate* process, but the combined effect of enacting their connections upon the same class instance depends on the order in which they were activated upon the model.

Table 7. Scenarios that affect a developer's error generation rate

```
SCENARIO ReducesErrGen ProjectModel
{
   CONNECTION TheDeveloper Developer
   {
      PROC LearningFactor 0.1;
      AFFECT ErrorGenRate ErrorGenRate - LearningRate;
   };
};

SCENARIO RaisesErrGen ProjectModel
{
   CONNECTION TheDeveloper Developer
   {
      PROC PressureFactor 0.2;
      AFFECT ErrorGenRate ErrorGenRate * (1 + PressureFactor);
   };
};
```

Consider that both connections in the *ReducesErrGen* and *RaisesErrGen* scenarios were enacted upon the same class instance. The resulting *ErrorGenRate* equation, reduced by the first scenario then amplified by the second scenario, would be described as:

ErrorGenRate = (ErrorGenRate - LearningFactor) * (1 + PressureFactor)

However, if the scenario activation ordering changes, enacting the *RaisesErrGen* connection before the *ReducesErrGen* connection upon the same instance, the *ErrorGenRate* equation would be changed to:

ErrorGenRate = (ErrorGenRate * (1 + PressureFactor)) - LearningFactor

In the second activation order, the schedule pressure effects are perceived prior to the new development technique effects. Depending on property values (learning and pressure factor), these two equations would show distinct behavior in a specific model. Thus, scenario ordering must be considered when connections from several scenarios are enacted upon the same class instance.

Scenario activation ordering is defined per class instance in a specific model. Since the modeler must indicate which scenarios are activated for every model instance, the order in which these activations are listed in the model will be preserved when activation ordering is involved.

7. Final Considerations and Outlook

This chapter described a modeling and a simulation process for system dynamics that allows the development of domain models and their specialization for particular problems. A domain model describes the classes of elements within the domain, detailing their properties, behavior, and relationships. The model represents a particular problem, describing the instances of each class that participates in the problem and defining their property values and associated instances. The model is built using a language closer to the domain concepts, while the domain model uses system dynamics constructors. We believe that, since the model construction language is closer to the user knowledge, it helps model development.

After describing the system dynamics metamodel, we have presented scenario models, an extension to the previous metamodel that allows the separation of uncertain assumptions from facts expressed in a model. Uncertain assumptions are described in separate models, namely scenario models, which can be activated upon a specific model. Such activation adjusts the original model equations to the formulations that describe the scenario, allowing a modeler to evaluate the impact of the scenario upon the model behavior. Scenarios allow modelers to extend the behavior of a model without direct and error-prone intervention in its equations.

We believe that a major advantage of the meta-modeling approach is the simplified process for creating specific project models, given a domain model. A strong limitation is that domain model development still depends solely on basic system dynamics constructors. While the distribution of equations among domain classes may help a modeler to concentrate on the behavior of one class at a time, it brings with it some difficulties regarding communication among classes and the public interface of stocks, rates, and processes that one class offers to the remaining system. Similar problems also apply to scenario models, whose construction demands knowledge of the domain model and the internal behavior of its classes.

Currently, we have a library of approximately twelve scenario models developed for the software project domain. These scenarios include theories regarding developers' productivity and error generation rates due to their experience, developers' productivity due to learning the application domain, effects of overworking and exhaustion upon developers, communication overhead, error propagation across the activities within a work breakdown structure, bad fixes, among others. We expect to create more scenario models for the software project management domain and use the proposed techniques as a training tool.

We have built a compiler that translates the metamodel representation and scenario model activations to traditional system dynamics constructors in order to allow model simulation. This tool is available, along with some scenario model examples and an extensive software project model, at the project website[2].

We have also conducted experimental studies[12] that yielded some positive results about the usefulness of scenario models in supporting decision making for software project managers. However, these studies also indicate that managers have difficulties in interpreting simulation results, so mechanisms to enhance the presentation of such results are needed. The outlook for this work includes the development of graphical tools to support the creation and evolution of specific models and scenario models. Such tools would be useful as a simulation environment, where a modeler could select scenarios and easily activate them upon a specific model developed for a domain.

Some limitations of the traditional system dynamics modeling paradigm can still be found in the metamodel representation. A strong limitation is related to system structure: though system dynamics models provide dynamic behavior, they rely on static relationships among the elements within a problem domain. Thus, model structure does not usually evolve over time. We are working on the concept of events, which would allow an analyst to influence model structure during a simulation run. Events would allow, for instance, an analyst to change the developer assigned to an activity without the need to rerun preceding simulation steps. We believe that this capability will bring simulation

---

[2] http://www.uniriotec.br/ ~marcio.barros/systemdynamics.html

closer to real-world situations, where managers decide on-the-fly to change the relationships within project environments (model structure).

We have used scenario models within a scenario-based project management paradigm[13], which proposes that a manager should plan and document the expected behavior of a software project as a system dynamics model. Since project behavior can be affected by unexpected events, management actions and strategies, the manager should test its sensitivity to combinations of such elements, evaluating the impact of these risks and whether they can challenge project success. Project management scenario models support decision-making by providing a library of known behavior models for management strategies and theories about developer's behavior that the manager can integrate into the baseline model describing the project to be developed. Scenario models can also describe optional process activities (such as inspections, specific configuration management techniques, formal reviews, and so on) that can be included in the project process to improve specific aspects (reliability, reworking reduction, artifact availability, among others).

In an industrial setting, where the scenario-based project management paradigm can be used to manage real software projects, senior managers should develop scenario models expressing experiences that they have collected by participating in several projects. These scenarios would allow less experienced managers to share senior managers' knowledge. In an academic setting, scenarios developed by experts and according to research presented in the technical literature could support training activities: students should use scenario integration and simulation to evaluate the impact of their decisions upon project behavior (such as cost, schedule, quality, and so on). Such an experiential environment may save students from repeating the same errors that they have already learnt from the simulator in real, industrial projects.

Acknowledgements

The authors would like to thank CNPq and CAPES for their financial investment in this work.

References

1. Abdel-Hamid, T., Madnick, S.E. 1991. Software Project Dynamics: an Integrated Approach, Prentice-Hall Software Series, Englewood Cliffs, New Jersey.
2. Tvedt JD. 1996. An Extensible Model for Evaluating the Impact of Process Improvements on Software Development Cycle Time, D.Sc. Dissertation, Arizona State University, Temple, AZ.
3. Lin, C.Y., Abdel-Hamid, T., Sherif, J.S. 1997. "Software-Engineering Process Simulation Model (SEPS)", Journal of Systems and Software, Vol. 38, Issue 3, September, pp. 263-277, Elsevier.
4. Madachy, R.J., Tarbet, D. 2000. "Case studies in software process modeling with system dynamics", Software Process: Improvement and Practice, Vol. 5, Issues 2-3, pp. 133-146.
5. Rai, V.K., Mahanty, B. 2001. "Dynamics of Schedule Pressure in Software Projects" , IN: The Proceedings of the 20th International Conference of the System Dynamics Society, Palermo, Italy.
6. Forrester, J.W. 1961. Industrial Dynamics, Cambridge, MA: The M.I.T. Press.
7. Booch, G., Rumbaugh, J. Jacobson I., 1999. The Unified Modeling Language User Guide, Object Technology Series, Addison Wesley Longman, Inc, Reading, MA.
8. Neighbors, J. 1981. "Software Construction Using Components", Ph.D. Thesis, University of California, Irvine, USA.
9. Barros, M.O. 2001. HECTOR - Metamodel Compiler to System Dynamics, available at URL http://www.cos.ufrj.br/~marcio/Hector.html (last accessed in 10/01/2003).
10. Barros, M.O. 2001. ILLIUM - System Dynamics Simulator, ILLIUM tool homepage at URL http://www.cos.ufrj.br/~marcio/Illium.html (last accessed in 10/01/2003).
11. De Marco T. 1982. Controlling Software Projects, Yourdon Press, Inc., New York.
12. Barros MO, Werner CML, Travassos GH. 2004. "System Dynamics Extension Modules for Software Process Modeling", ProSim'03, 2003 Workshop on Software Process Modeling, Portland, OR, USA.
13. Barros MO, Werner CML, Travassos GH. 2004. "Supporting Risk Analysis on Software Projects", The Journal of Systems and Software, Vol. 70, Issues 1-2, February, pp. 21-35, Elsevier.

Chapter 5

PEOPLE-ORIENTED CAPTURE, DISPLAY, AND USE OF PROCESS INFORMATION

Jens Heidrich, Jürgen Münch, William Riddle, Dieter Rombach

Fraunhofer Institute for Experimental Software Engineering
Sauerwiesen 6, 67661 Kaiserslautern, Germany
E-mail: {heidrich, muench, riddle, rombach}@iese.fraunhofer.de

Project success demands that process performers have accurate, up-to-date information about the activities they should perform, any constraints upon activity performance, and guidance about how to effectively and efficiently perform their activities. The goal of this chapter is to describe support for people-oriented capture, display, and use of process information that experience has shown is highly beneficial. The chapter reviews several state-of-the-art approaches for supporting people-oriented process performance, illustrates challenges of providing this support, and presents experience from practice. We describe different kinds of process knowledge and discuss a method for collecting one kind of process knowledge – measurement data – in a goal-oriented way. We present different ways to display process information in order to satisfy information needs of people involved in a software development project, including the generation of process documentation, role-based workspaces, and control centers for software development. Furthermore, we illustrate how process information can be used to support process performance through the use of not only workspaces and control centers but also process enactment and experience management. The approaches presented in this chapter can be seen as a contribution towards supporting people-oriented software development.

1 Introduction

Well-designed, accurately-performed processes are critical to the successful conduct of an organization's projects. This is particularly true when developing large systems by carrying out many, highly interdependent, activities. Some activities are quite simple but others, such as project planning, coordination, cooperation, control, and improvement, may be quite complex. Some activities may be "algorithmic" (defined by concrete procedures) and some may be "creative" (stochastically undetermined). Some may be performed by teams of developers; others may be enacted by machines. All activities must contribute to meeting the project goals which will vary across projects.

As a result, the processes must be designed for effective, efficient, and accurate performance in a wide variety of contexts. Three particularly difficult challenges are:

- The conduct of development processes cannot be completely automated because performance is largely human-based and, therefore, has to deal with non-deterministic behavior of human process performers. This often causes problems such as non-predictable deviations or confusion regarding roles and responsibilities.
- Contexts characterize a project's environment and consist of organizational, technical, and personal characteristics that influence the project's processes and their performance. The relationship of the context and the processes is often quite hard to understand, especially for the development of large-scale, complex systems. This makes it difficult for process performers to obtain the necessary or appropriate information and assess the impacts of different contexts.
- The contexts vary between projects and even within projects. Because development activities are context-dependent, the processes need to be adapted to different contexts and this adaptation is often quite difficult. In addition, the processes may need to be adapted "on the fly" during project performance. Rapid context switches often lead to unnecessary rework when context switches are insufficiently supported.

The development of large and complex systems that may include hundreds of hardware and software components is quite complex because of these challenges. Other complexities occur in small, team-based development projects. Many problems are related to the fact that software development is largely human-based. People need help in meeting these challenges. This includes customizing process information to specific needs for different contexts. Necessary process support includes improved communication, detailed reasoning about process features, guiding people when performing processes, improving both processes themselves and their results, and automating process steps to gain deterministic process behavior[1, 2]. People-oriented process support requires managing large and complex processes contexts by reducing the cognitive load, supporting long-living processes by providing mechanisms for context switching, and supporting collaboration by applying team-based and multi-disciplinary approaches.

The goal of this chapter is to describe support for people-oriented capture, display and use of process information that experience has shown is highly beneficial. The chapter reviews several state-of-the-art approaches for supporting people-oriented process performance, illustrates the challenges of providing this support, and presents experience from practice. Most of the approaches and examples stem from the software engineering domain; i.e., the processes are "software development processes". However, much of the material in this chapter can be applied to "development processes" from other domains (such as development processes for mechanical systems) or other – business-oriented – process domains within an organization, such as marketing and sales processes.

1.1 General Concepts

People-oriented process support requires mechanisms supporting the interface between processes and people; this is, in essence, the focus of this chapter. The concepts of *role* and *agent* are fundamental to discussing these mechanisms and are therefore explained in this sub-section.

A role definition indicates the role's purpose with respect to the process. In essence, role definitions define the parts that people play as they participate in carrying out the process. Role definitions are analogous to the part definitions found in a script for a theatrical production. An important aspect is that the role definition is specific to the process. If role definitions in different process descriptions have the same name, then this is incidental. The purpose, scope, and nature of the role are solely as specified in its definition for the specific process.

Just as people play the parts in a theatrical production, people play roles in process performance. To specify that a person is playing a role, we say the person *occupies* the role. Further, we do not talk in terms of specific people (Bob, Michele, or Pete), but rather talk in terms of agents.

The net effect is that there are two levels of abstraction reflecting two very useful separations of concern. The role vs. agent abstraction allows the separation of a concern for what a person must do when participating in a particular process from a concern for what the person must be able to do as an employee who may be assigned to participate in various processes. This separation of concern is highlighted by noting the difference between a role definition, which is process specific, and a job description, which is position specific and usually fairly neutral about the specific processes the employee will be assigned to.

The agent vs. person abstraction allows a separation of concern for a general spectrum of qualifications, abilities, and experiences — as typically indicated in a job description — from a person's specific qualifications, abilities, and experiences. This abstraction has two important side-effects. First, it allows a group of people to share a particular position. An example is a `System Administrator` position which is often filled by assigning a group of people who each work part-time in this position. The other side-effect is that it becomes possible to think of filling a particular position with a tool, effectively automating the position. An example is a `Request Filter` agent that filters incoming requests into several different "bins," a capability that is necessary for a variety of processes and may quite often be automated.

A role definition must treat three different aspects of the role as part of a process:

- Responsibilities: the role's obligations and permissions with respect to the process; for example `create financial report`, `assure project success`, and `can access employee records`.
- Activities: the role's participation in activities, perhaps accompanied by time sequencing information; for example, `gather the requirements` and `develop the design`.

- Abilities: skill and experience requirements for agents who occupy the role; for example, `trained in using Word` and `familiar with the Delphi approach to brainstorming`[a].

These three aspects of a role correspond to three different approaches to supporting people during process performance. All three approaches allow agents to find and focus on an activity they must carry out; the three approaches vary with respect to the view presented to the agent as a starting point. In a responsibility-based approach the agent starts with a view that emphasizes its responsibilities. In an activity-based approach, however, the starting point reflects a collection of inter-related activities and the agent selects one of the activities that the role participates in. Finally, in an ability-based approach, agents start with a view indicating the capabilities needed to successfully occupy a role and the agent may select an activity that requires a specific capability. In this paper, we focus on responsibility and activity-based approaches.

People-oriented process support needs to consider the human characteristics of individuals and teams. These characteristics comprise motivation (i.e., the stimulus a person has about achieving some result), satisfaction (i.e., the fulfillment of a need or the source or means of enjoyment), skills (i.e., knowledge and training), and experience (i.e., practical knowledge and experience resulting from observing or participating in a particular activity).

Process support should be customized to human characteristics and it should recognize that it can influence these characteristics. The motivation of a highly experienced developer, for example, could be decreased by prescribing finely detailed approaches to carrying out activities. The skills of an inexperienced developer could be amplified by providing detailed guidance for an activity.

Human characteristics may have a major influence on the results of a development project. The following empirical findings and hypotheses

---

[a] These statements could appear in a job description to specify criteria for evaluating potential employees. Here, they are being used to specify the criteria for some agent assigned to the role. These are analogous uses, but the first concerns an agent's general skills and experience whereas the second concerns the skills and experience needed for a specific process.

(discussed in "A Handbook of Software and Systems Engineering"[3]) are related to skill, motivation, and satisfaction and can be applied for designing people-oriented process support. According to Kupfmüller's law, humans receive most information visually. It is important to consider that not all senses are involved equally during the reception of process information. The predominance of the visual system could be exploited for displaying process information via pictures, graphs, or signs. Krause's law states that multimodal information is easier to remember than single-mode information. Including multimedia capabilities in process support could utilize this finding. Miller's law, defined by the psychologist George Miller, says that short-term memory is limited to 7 +/- 2 chunks of information. This indicates how much the complexity of processes and contexts needs to be reduced for displaying them adequately to process performers. Many other empirical findings should be considered when developing people-oriented process support, for example: human needs and desires are strictly prioritized (Maslow-Herzberg's law); motivation requires integration and participation (McGregor's hypothesis); and group behavior depends on the level of attention (Hawthorn effect).

These empirical findings and hypotheses should be carefully considered when providing people-oriented process support because adherence or non-adherence to them is a major determiner of success or failure.

1.2 Contents

The chapter is organized into three parts regarding capturing process information, its display, and its use. Section 2 addresses the collection of different kinds of process knowledge and discusses the Goal Question Metric (GQM) paradigm as a goal-oriented method for collecting process-related measurement data. Section 3 deals with the different ways process information may be displayed in order to satisfy process performer information needs. In the section, we discuss the generation of process documentation, role-based workspaces, and control centers for software development. Section 4 illustrates how process information can be used during process performance. In the section, we discuss the importance of not only workspaces and control centers but also process enactment support. In the section, we also discuss experience management issues; that is, various ways in which process information can be used to improve future process performance. In each section, we discuss related work and future trends as appropriate. Finally, section 5 gives a brief summary of the material in this chapter.

Fig. 1 gives an overview of the sections in terms of their relationship to a simple development process and its major roles.

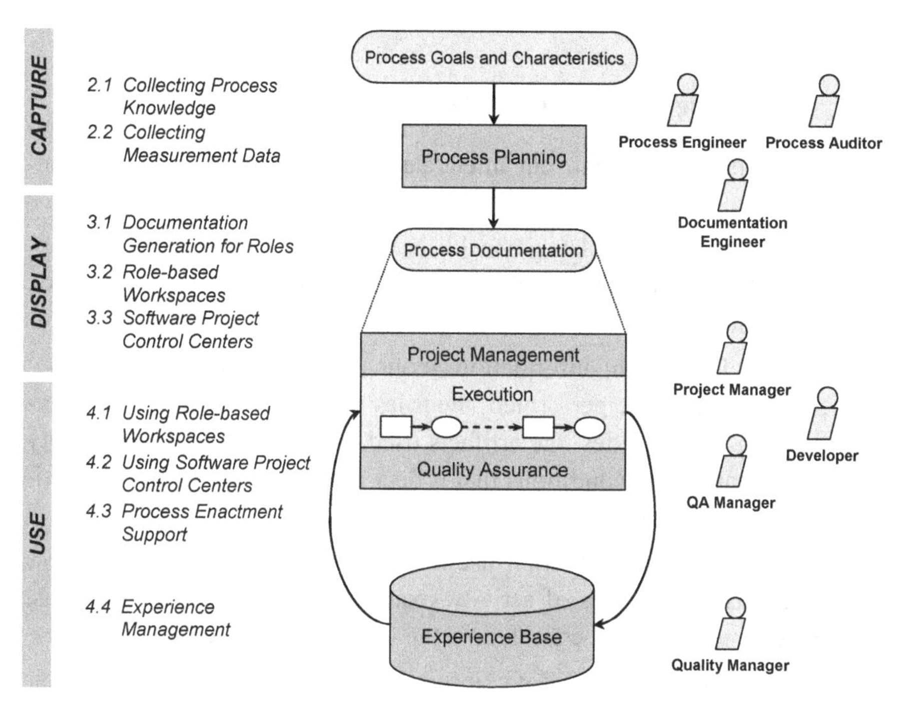

Fig. 1. Section Focus with respect to a Simple Development Model (not addressing all Planning and Enactment Issues).

2 Capturing Process Information

Access to information about a process is critical to its effective, efficient and accurate performance. This information comes from many sources: process designers, process performers, customers and other stakeholders concerned with the outcome, and corporate executives interested in the degree to which the processes support achieving the organization's business objectives. Once collected, the information is typically organized into process handbooks, with different handbooks rendering the information as needed for particular roles (for example, a `Developer Handbook` and a `Process Auditor Handbook`).

In this section, we the first define various kinds of process information. After briefly discussing the collection of these kinds of process knowledge, we then address the collection of measurement data (one kind of process knowledge) in more detail.

2.1 Collecting Process Knowledge

There are many different, but inter-related, kinds of process-related information:

- Process Assets: templates, checklists, examples, best-practice descriptions, policy documents, standards definitions, lessons-learned, etc., useful during process performance.
- Process Definition: information about a specific process including: the activities to be performed, the roles that agents occupy when performing the activities, the artifacts used and produced during activity performance, the conditions that reflect progress, and the assets pertinent to the process.
- Process Status: information useful for controlling or tracing process performance in terms of activity sequencing, role/agent assignments, the degree of artifact completion, the usage of assets, etc.
- Process Measurement Data: information characterizing a particular performance of a process in terms of the expended effort, performance of the personnel, resource utilization, etc.
- Project-specific Data: information created during a process performance including: specific documents created and worked on during the project, records of group meetings, white papers discussing various design options, lessons learned, etc.

In the following sections we discuss the collection of each of these kinds of process knowledge.

2.1.1 Process Assets

Process assets may be collected from several sources. Standardization efforts[4] not only define the standard but also provide example templates and checklists as well as examples of good (and bad) uses of these assets. Definitions of Process Maturity Frameworks, for example the CMM[5], are also often the source of templates, checklists, and examples of good or bad artifacts. Framework-oriented sets of assets are also commercially available (for example from pragma[6] and the FAA[7]).

Corporate-wide process definition groups are normally responsible for developing assets supporting the organization's processes. This primarily includes templates, examples and checklists. It also includes the definition of best practices as they should be practiced within the organization.

These process definition groups are also frequently responsible for assuring that the organization's process assets are influenced by the experience gained during project performance. This includes collecting new templates, checklists, examples, etc. It also includes collecting lessons learned. Finally, it includes updating existing assets to reflect experience in using them.

Many organizations have devoted considerable effort to collecting their assets, and making them readily available, in a corporate-wide Process Asset Library (PAL). A PAL organizes the assets according to various cataloguing dimensions. These dimensions reflect the assets' general characteristics (their general types, the application programs needed to use them, reactions stemming from their use, their pertinence to various types of projects, etc.). An asset's pertinence to specific processes may also be used as a cataloguing dimension. In the following, however, we indicate that process definitions may provide better, role-specific as well as process-specific, "doorways" into a PAL.

2.1.2 Process Definition

Information about a process may be collected in several ways. Typical alternatives are observing real projects, describing intended activities, studying the literature and industry reports, and interviewing people involved in a project. These approaches to gathering process knowledge are varied and may be combined in many ways. All of the various ways result, however, in one of two distinctly different kinds of process definitions: *descriptive process models* and *prescriptive process models*.

Descriptive process models describe processes as they take place in real projects, i.e., descriptive process models are the result of observation. Descriptive models describe how a particular software system was developed, are specific to the actual process used to develop the system, and may be generalized only through systematic comparative analysis[8]. They can be used for analysis purposes and are often the basis for understanding and improving existing practices. Descriptive process models can be used as a baseline for improvement programs and they can be a basis for identifying strength and weaknesses. Descriptive process modeling is a good means for obtaining a process description that accurately describes the real process. This helps gain a deeper understanding of the process, a prerequisite for effective process measurement and assessment.

For conducting descriptive modeling, Becker, Hamann, and Verlage[9] propose a two-phase procedure consisting of eight steps. The set-up phase comprises the configuration of the process modeling approach. The steps in this phase are performed relatively infrequently. The execution phase concerns a particular use of the process. All steps in the execution phase should be performed for each inspection of a particular use. An enumeration of the phases and steps is:

- Phase I: "Set-up"
  - Step 1: State objectives and scope.
  - Step 2: Select or develop a process modeling schema.
  - Step 3: Select (a set of) process modeling formalisms.
  - Step 4: Select or tailor tools.
- Phase II: "Execution"
  - Step 5: Elicit process knowledge.
  - Step 6: Create model.
  - Step 7: Analyze the process model.

o Step 8: Analyze the process.

One approach to descriptive modeling, regardless of how it is conducted, is *multi-view modeling* (MVM)[10]. In this approach, the underlying idea is to reduce process description complexity by eliciting role-oriented process information (e.g., from the point of view of a `Tester` or a `Requirements Analyst`) because people are unable to look at complex processes as a whole and usually focus on information relevant only to their roles. Because roles collaborate, there is typically an overlap of process information in different role-oriented views. Experience has shown that different people occupying different roles in a project often have a varying understanding of the overall process. MVM permits information from different role-oriented views to be integrated to develop a consistent, comprehensive descriptive process model.

MVM supports the elicitation of role-oriented process models, the analysis of similarities and differences between these models, the analysis of consistency between different models, and finally the integration of various views into a comprehensive process model. Benefits of this approach are: it considers the perspective and experience of individual persons or teams, it supports role-oriented modularization of complex process models, it supports the identification of inconsistencies and inefficiencies between role-specific process views, and it allows the reuse of role-specific views.

In contrast to a descriptive process model, a prescriptive process model defines guidelines, frameworks and details for a new approach to carrying out the process (Scacchi[8]). It specifies how activities should be performed, and in what order. It also establishes conditions that can be used to guide or track process performance; this is discussed in the very next section. Finally, it establishes contexts in which process performers may effectively, efficiently and accurately satisfy the responsibilities for the roles that they occupy; this is discussed in section 3.2.

2.1.3 Process Status

Process status information characterizes the state of process performance. The fundamental notion is *process element state*: a measurable attribute of some process element. For example, an artifact's state might be drafted, approved, or rejected. As another example, the state of an activity might be suspended, active, or completed. A statement about a process state that can be used to control the process or characterize progress is called a *condition*: a Boolean expression referring to the states of process elements. An example condition is Design Document approved and Coding Standards identified. The validity of a condition changes over time as a result of *events*: actions that establish new states for process elements (or perhaps merely re-establish their existing states). An example event is design review finishes which might establish the state of a Design Document as either approved or rejected and thereby affect the validity of conditions defined with respect to the state of the Design Document.

Status information may be collected by the techniques identified in previous sections (for example, by interviewing process performers). The most important purpose is to collect information about the possible states for process elements (for example, the possible states for a Financial Report artifact). This is often simplified by identifying states for types of process elements, for example, by indicating that all artifacts may have the states drafted, approved, or rejected. Another purpose of status information collection is to identify the conditions needed to assess progress and activity pre- and post-conditions useful for controlling process performance. A third purpose is to identify the events that change process element states and affect the validity of conditions; the events are normally strongly tied to points in the definition of an activity at which the events occur.

Once collected and used as part of a prescriptive process definition, status information may guide the collection of data about the actual status at various points during process performance in order to check that the actual performance matches the intended performance. Collection of actual status information during process performance requires instrumentation of the performance-time support system. Amadeus[11] is an example of a system providing this instrumentation; it uses scripts to control data collection and these scripts can be developed using the status information in a prescriptive process description.

The status information in a prescriptive process definition may be used to control, rather than merely track, process performance. This use of status information to support process enactment is discussed is section 4.3.

2.1.4 Process Measurement Data

Process measurement data are collected during process performance and characterize special aspects of the process, related products, and resources, especially quality-related aspects. Example quality-related aspects include: the effort in person-hours a certain developer spent on a certain part of the process, the performance in lines of code per hour of the coding personnel, or tool usage in hours. Process measurement data provides quantitative information about process performance to complement the qualitative status information.

While project status information is collected in order to track or control the state of the overall project, measurement data are usually collected in order to control certain quality aspects. A project manager would basically be interested in the former and a quality assurance manager in the latter. Therefore, status information and measurement data provide qualitatively different, but complementary, views of process performance.

To control a software development project, it is crucial to know what measurement data have to be collected relative to defined measurement goals of the project. We address this issue in more detail in section 2.2.

2.1.5 Project-specific Data

Project-specific data are created by carrying out the process. For example, performing a `Create Design` task will create a `Design Document` that is a specific instance of the `Design Document` artifact specified in the process description. Project-specific data are collected during process performance as the agents occupying the process' various roles create and modify the specific artifacts produced during process performance.

Project-specific data provide a concrete record of the effects of a process performance, complementing the status information and measurement data characterizing how these effects were obtained. Most importantly, this includes the intended results of the process. It also includes other documents such as records of project meetings and white papers discussing various design options.

2.2 Collecting Measurement Data

Measurement data are needed to analyze quality aspects of a project, the software development process used by the project, the products it produces, and the resources it uses[12]. Examples of measurement data are: the effort of the process `Create Requirements` in person-hours, the complexity of the product `Requirements Document` in function points, or the cost conformance of the project `Building Automation System` in US dollars above or below the planned costs.

But how can we decide what to measure and how to interpret the collected measurement data? Basically, we may distinguish two types of measurement approaches. The first one starts with measurable observations and relates them to measurement objectives and goals. We call this approach bottom-up because it starts with concrete measures and ends up with abstract measurement goals. The second type starts with the definition of a measurement goal and derives concrete measures from it. We call this approach top-down because every measure is derived and interpreted in the context of a specific goal. Starting with a measurement goal eases the development of adequate, consistent, and complete measurement plans.

One top-down approach is the Goal Question Metric (GQM) paradigm developed by Victor Basili and David Weiss[13]. Its main idea is to define a measurement goal and systematically derive questions and finally metrics. The approach can be applied to software processes, products, and resources and gives guidance on how to derive metrics in a goal-oriented way.

Before describing this approach in more detail, it is important to define some basic concepts:

- A *software entity* is a specific process, product, or resource pertaining to a software development project for which measurement data are being collected. For example, `Create Requirements`, `Create Design`, `Coding`, and `Testing` are process entities.

- An *attribute* is a characteristic of a specific entity. For example, the length of the product `Code` is a product-related attribute.
- A *metric* is a (numerical or enumerated) scale that characterizes an attribute of a specific entity. For example, the length of product `Code` can be measured as lines of code (including or excluding comments).
- *Measurement* is the process of assigning certain values (defined by the metrics) to attributes of software entities.
- *Measurement data* is the set of concrete data for all metrics measured during the performance of a software development project.

Some metrics can be measured directly, whereas others have to be computed from other metrics. We call the first *direct* and the latter *indirect metrics*. For example, the indirect metric `simple design complexity` for an object-oriented system may be computed from the direct metrics `number of classes` and the `number of relationships among classes`.

2.2.1 *Definition of GQM Plans*

The first step in setting up a measurement plan using the GQM paradigm is to define all of the goals for a measurement program. Each GQM goal definition consists of five different components[14]:

- The *object* defines the central node for measurement, namely the object we want to analyze. For example, a certain process, product, or resource, or even the overall project may be a legal measurement object.
- The *purpose* describes the intention we have in setting up a measurement plan. For example, we want to characterize, improve, control, predict, or simply analyze the measurement object.
- The *quality focus* defines the characteristics of the analyzed object we are interested in. For example, reliability, usability, security, safety, scalability, performance, efficiency, or maintainability of the measurement object.

- The *viewpoint* describes the perspective from which a quality aspect is analyzed. For example, a developer, the project manager, the quality assurance manager, or the customer. This component is especially important to supporting people-oriented measurement; it helps in identifying the different groups of people interested in the collected measurement data and in avoiding needless data collection.
- The *context* defines the environment where measurement takes place. Usually, measurement data highly depend upon the context they are collected in, and the results that are obtained are not transferable to different environments. For example, effort data originating from a development project for embedded systems is probably not transferable to web-based application development.

An example of a measurement goal definition is: Analyze *the inspection process* for the purpose of *improving* with respect to *efficiency* from the viewpoint of *the quality assurance manager* in the context of *company A's implementing the automation system BAS 2004*.

The second step in setting up a GQM-based measurement plan is to derive questions for each measurement goal. Answering these questions should support assessing the measurement goal. We distinguish between two types of questions: Questions regarding the quality focus of the measurement goal and questions regarding variation factors, i.e., factors that influence the values of the quality focus measurements. The latter type of question can further be divided into questions regarding processes (for example, process conformance, domain understanding, and so on) and questions regarding products (for example, logical and physical attributes of products such as size and complexity, development costs, and changes). Questions illustrating a quality focus are:

- How many defects are in the requirements document?
- What is the distribution of the defects with respect to a set of defect classes?
- How many defects were found when inspecting the requirements document?
- How much does it cost to fix all the defects found in the requirement document?

On the other hand, questions that influence the quality focus measurements and therefore address variation factors are:

- Process-related: How experienced are the developers? What kind of inspection technique is used?
- Product-related: What is the complexity of the requirements document?

If a question is too complex to be answered directly, we may refine it to a set of simpler, more specific, questions. For example, we can refine the question regarding requirements document complexity to the following three, very specific, questions: How many functional, non-functional, and inverse requirements and design decisions are listed? How many use cases are included? The composition of the answers to the more specific questions leads to answering the original, complex, question.

After defining all of the measurement goal-related questions (and refining them), we are able to derive some metrics that help in answering the questions. This is the third step of the GQM approach. We can define more than one metric for a single GQM question. For example, regarding the question concerning developer experience, we can measure the number of years employed for each developer, the general degree of experience (e.g., `high`, `medium`, `low`), or the developer's experience with a certain requirements analysis tool. As mentioned, each metric has values along a certain scale and the scale indicates which values may be combined, and in which ways, to get valid results. For example, we may add the rework effort for every developer of the requirements document (because the values are on a rational scale), while values of the general degree of developer experience may not be added (because they belong to an ordinal scale and the result of adding `high` and `low` is not defined by the scale).

An important question is, "How can we find a set of relevant measurement goals and derive questions and metrics?" An enhancement of GQM uses *abstraction sheets*[15] to identify components of measurement goals and to analyze the quality focus and its variation factors in more detail. In essence, an abstraction sheet summarizes a GQM plan. An example is shown in Fig. 2.

Object	Purpose	Quality Focus	Viewpoint	Context
Requirements Inspection Process	Characterize	Efficiency	Quality Assurance Manager	company A's building automation system BAS 2004

Quality Focus	Variation Factors
QF1: Number of included defects QF2: Distribution according to defect classes QF3: Percentage found during inspection process QF4: Total rework effort in hours ...	VF1: Experience of developers VF2: Inspection type ...

Baseline Hypotheses	Impact of Variation Factors
QF1: 100 QF2: 40% omission, 30% ambiguous information, 20% incorrect fact, 10% miscellaneous QF3: 20% QF4: 1000 h ...	VF1 high => QF1 low => QF4 low VF2 perspective-based => QF3 high ...

Fig. 2. Sample of a GQM Abstraction Sheet.

As shown in the figure, an abstraction sheet consists of five different sections. The first section (at the top of the example sheet) gives an overview of all the components of the measurement goal. The second section identifies direct metrics related to the quality focus. Not all metrics defined in a GQM plan are represented on abstraction sheets, only the most important ones. The third section identifies variation factors; that is, metrics that influence the values of quality focus metrics. (These two sections appear in the center of the example sheet.) The fourth section defines predicted values for the metrics defined in the quality focus section; that is, it identifies expected values hypothesized before collecting the real measurement data. These data can be used to control the measured quality aspects of the project during project execution. The fifth section indicates the expected impact of variation factors for the metrics identified in the quality focus section. (These two sections appear at the bottom of the example sheet.) An abstraction sheet is usually constructed during structured interview sessions with all of the measurement program's stakeholders. The GQM viewpoint helps in identifying the interests of different project (and organization) roles and in inter-relating different measurement needs.

2.2.2 Application of Measurement Plans

After defining the measurement plan (e.g., a GQM plan), the next step is to select techniques and methods for collecting the measurement data. This includes assigning persons responsible for data collection and validation. The tasks that are necessary to perform data collection are called *data collection procedures*. Van Solingen and Berghout[12] suggest that at least the following questions must be addressed when setting up data collection procedures:

- Which person should collect which metric?
- When should a person collect a metric?
- How can measurement data be collected efficiently and effectively?
- To whom should the collected data be delivered?

We can distinguish between data collected automatically by a tool, such as lines of code or design complexity, and data that have to be collected from people, such as effort in person-hours for a certain process. Experience has shown that, in general, the most valuable information is that collected from people rather than by tool-based analysis.

Data can be collected from people through manual (paper-based) forms or electronically (e.g., via web-based forms, e-mail, or spreadsheet forms). The greatest advantage of an electronic data collection system is that the collected data may be used for project control purposes (if interpretation and analysis of the raw measurement data can be automated).

Data can be collected when special events occur (such as when a project milestone is reached, at the end of major activities, or upon completion of important products) or continuously at periodic time points (e.g., every day or once a week). The best collection strategy to use for a metric depends on several factors, such as the collection approach (automatic versus manual data collection), the metric being measured, and the people entering the measurement data.

3 Displaying Process Information

The information needs of the people involved in a software development project varies according to the roles they occupy. A `Project Manager`, for example, needs different process information than a `Developer`. The former will be interested in the state of the overall project (e.g., a list of all uncompleted tasks), while the latter is most interested in process information regarding specific development activities (e.g., how a certain activity has to be performed). The goal of this section is to illustrate a variety of ways process information can be presented and how different project roles may benefit from different presentations. First, we describe the generation of process documentation in order to support different project roles. Secondly, we discuss role-based workspaces displaying process information in an organized, role-specific, manner. Finally, we describe control centers for software development as a systematic way for interpreting and visualizing measurement data.

3.1 Documentation Generation for Roles

Process documentation is generally thought of as information that helps process performers do the "right thing" at the "right time." As such it defines the work that should be done, dependencies among work packages assigned by the project manager, the responsibilities of the various roles, the conditions that control the sequencing of the work, the conditions that can be used to track progress, etc. Process documentation is most usually organized according to a logical decomposition of the work that must be done with chapters for major phases, sections for major activities within each phase, and sub-sections for specific tasks.

Process documentation content and organization reflect task-related questions posed by *project members*, agents who are assigned to a project and responsible for carrying out the process. Typical questions are: What do I have to do to perform a task? Which tasks am I responsible for? How can I determine whether or not I have successfully completed a task? Where can I find a particular template?

Across all the roles in a process there will be many questions of many different kinds. For each kind of question, there will be one or more views that help in answering questions of this kind. For example, an activity decomposition view helps in asking questions about the interdependencies among activities and an artifact lifecycle view helps in answering questions about artifact state sequences.[b]

The information needed to answer some role's questions is typically documented in handbooks. The use of handbooks for software development processes has been recognized widely as beneficial in order to perform systematic, traceable projects. Nevertheless, software developers generally face problems in using the handbooks typically provided for software development processes. The reasons for these problems include:

- The handbooks are lengthy, perhaps hundreds of pages, and often not very well structured, thus complicating information retrieval.
- The handbooks are frequently out-of-date. Because new versions are difficult and time-consuming to produce, they are infrequently developed.
- Formal notations may be used to achieve high degrees of precision; graphical representations may used to increase the understandability of descriptions using these notations. Process descriptions are, however, usually informal. Among other things, this makes it hard to customize the processes to match the characteristics and needs of a specific project.
- The dynamic behavior of the process is not well-specified, again because the descriptions are informal. Ambiguity is common, and different agents have different understandings of the behavior.
- The consistency, clarity, and completeness of informal software process descriptions cannot be easily ensured. Costly, lengthy reviews are therefore needed to assure consistent, clear, and complete handbooks. These reviews are frequently not done.

Even more problematic is that one handbook cannot conveniently meet the needs of all the roles having questions about the process and its

---

[b] In paper-based documentation, one view is, of necessity, used in the body of the documentation and other views are provided in appendices. In web-based documentation, all of the various views may be provided as "top-level doorways" into the process information.

performance. Many handbooks are needed, each oriented towards some role's domain of questions. This severely complicates the maintenance of an organization's process descriptions because changes and additions have to be accurately and consistently made to a possibly large number of handbooks.

The notion of Electronic Process Guides (EPGs)[16] was developed to address these problems. The general purpose of an EPG is to guide software developers in doing their tasks by providing fast access to the information they need (e.g., to activity and artifact descriptions, to assets such as checklists, etc.). As implied by its name, an EPG provides online, internet/intranet descriptions of the process. However, rather than merely being electronically provided versions of paper-based handbooks, EPGs are extensively hyper-linked to facilitate navigation through the information about the process. Moreover, EPGs are automatically generated from information captured, non-redundantly, in a process model. Because they are automatically generated, the maintenance of a consistent set of handbooks is much easier and less error prone – any change to the information in the process model is automatically reflected in all the handbooks (once they are regenerated). This makes it much more feasible to provide a set of handbooks with each essentially filtering the information and rendering it in a way most meaningful to some role's questions. Finally, the process model may be formal, rather than informal, and this introduces a rigor that not only precludes common errors (such as using two different names for a process element) but also enables checking the consistency and completeness of both the dynamics and the specification of the process. As a result, EPGs provide up-to-date, accurate, consistent and complete information about a process packaged into the different forms needed to support the different roles in a process. Experience has shown that this considerably enhances the efficiency and effectiveness of process performers.

To explain EPGs further, and indicate their value, we describe two tool suites which have been developed to provide an EPG capability. The first is the SPEARMINT®/EPG tool suite developed at the Fraunhofer Institute for Experimental Software Engineering (IESE)[17]. The second is the Process Management Capability (PMC) tool suite developed at TeraQuest Metrics (TQ)[18]. Both were designed to provide support for managing large process models and the automated generation of online documentation of process handbooks from these process models. They share many common characteristics, but differ in terms of some of the capabilities they provide. In the following, we first describe their common characteristics and then their key differences.

In both cases, the primary objectives were to: (1) support the development of a process model, (2) eliminate common errors, (3) support analysis of the completeness and consistency of both the process and its descriptions, and (4) support the generation of a handbook providing process performance guidance. The tool suites took similar approaches to meeting these primary objectives:

- A well-defined, Entity-Relationship-Attribute style, process modeling technique is used to capture process information. The process model reflects the elementary "information chunks" pertinent to many different question domains (for example, an information chunk that identifies all the roles participating in a task).
- A well-defined data-storage format is used to capture, non-redundantly, the basic "elemental facts" of which information chunks may be composed. For example, the fact that a specific role participates in a specific task would be stored as a single, elemental, fact although it may appear in many different information chunks.
- An editing tool allows a process engineer to capture and maintain the information about a process by establishing and modifying the elemental facts. The editor allows the process engineer to change a set of inter-related facts about the process in a well-coordinated way. The editor may also enforce process modeling rules, for example, the rule: `every task must have at least one participating role`. Finally, the editor allows some degree of customization of a process model through the definition of additional process-element attributes.
- A publishing tool supports the generation of a set of web pages which constitute a process handbook.
- A viewing tool supports the generation of reports (most often also in the form of web pages) supporting a process engineer's work. This

includes, for example, reports concerning inconsistencies and incompleteness.

The differences follow from the fact that the SPEARMINT®/EPG tool suite is focused on the process engineer's need to easily model and analyze a process during its development whereas the PMC tool suite is focused on the process engineer's need to customize an EPG's look-and-feel to the needs and desires of a particular organization. As a result, in the SPEARMINT®/EPG tool suite:

- The editing tool uses a graphical notation for describing processes. It distinguishes the following types of process elements: activities, artifacts, roles, and tools. The notation allows graphically denoting relationships among the process elements as well as the attributes defining the elements' measurable characteristics.
- The viewing tool supports a wide variety of software development and process engineering views. These were defined specifically to support distributed process planning by providing the appropriate representations for reviews.

In the PMC tool suite, however:

- The process model (and editing tool) may be customized to an organization's specific process architecture, i.e., the notions used to describe the organization's processes. The tool suite is based on the Collaborative Process Enactment (COPE)[19] generic process architecture which defines process entity categories (activities, roles, artifacts, conditions and assets) as well as some basic attributes and relationships. It may be customized to reflect an organization's process entity types, their attributes and their relationships.
- The publishing tool is controlled by templates for the various kinds of web pages that appear in a generated handbook. Each template describes the format for the page and identifies the elementary facts that need to be retrieved to generate an instance of the page. A template also describes the computation needed to infer information — for example, information about the flow of artifacts among tasks — from the elementary facts. Multiple sets of templates could be used to gen-

erate multiple handbooks differing either in their look-and-feel or in their role orientation.

Taken together, the two tool suites have been used to create web-based documentation for more than two dozen processes (examples are discussed in several papers[20, 21, 22, 23]; additional examples may be found at *http://www.iese.fhg.de/vincent/examples*). Most of these have been software development processes; a few have concerned an organization's business processes. The experience gained in preparing these EPGs has indicated that both the modeling/analysis enhancements provided by SPEARMINT®/EPG and the EPG-customization enhancements provided by PMC are necessary (and, as a result, the two tool suites are currently being integrated). The experience has also demonstrated the benefits of the EPG approach to process documentation: processes may be developed more rapidly; common errors may be precluded; consistency and completeness may be more extensively and accurately verified; and the time/effort required to maintain and deploy accurate, up-to-date process descriptions is reduced.

Two additional, potential advantages have been noted but not yet fully realized in either tool suite. First, process models are an appropriate means for storing software development knowledge. In general, reusing experience is a key to systematic and disciplined software engineering. Although there are some successful approaches to software product reuse (e.g., class libraries), all kinds of software-related experience, especially process-related experience, should be reused. Process models can be used to capture process-related experience, and this experience can be stored using various structures for an experience repository (e.g., type hierarchies, clusters of domain specific assets). Second, an EPG-based approach to process development allows several kinds of (automated) analyses, which may be performed before the project starts, during process performance and in a post-mortem fashion after project termination. Process models can, for example, be analyzed statically (e.g., to check for consistency) or dynamically (e.g., to check the flow of artifacts across their interface). The latter is important during the modeling of the interfaces of distributed processes.

3.2 Role-based Workspaces

Process handbooks, whether they are created by hand or by using an EPG-based approach, certainly provide the ability for an agent to effectively, efficiently and accurately occupy an assigned role. The major reason is that they contain the information that agents need to answer questions they have before starting to work on, or while working on, their assigned tasks.[c]

Agents also need the ability to focus their attention on the information pertinent to a specific task or a specific question. Locating and organizing the needed information can be extremely difficult no matter how well a role-specific process handbook is designed. In this sub-section, we introduce an approach for focusing on the information pertinent to a specific task. The aim of this is to effectively support project members. When focusing on supporting project managers, a different approach is needed. We discuss this different approach in more detail in section 3.3, when talking about project control.

3.2.1 Role-based Workspace Purpose

We define a *role-based workspace* to be a working environment supporting an agent occupying a role during process performance. A workspace has three major intents. One is to provide access to the documents, tools and assets the agent needs to perform tasks. The second is to help the agent assure that his/her work is satisfactory. The third is to facilitate collaboration with other agents occupying other roles. The first two are discussed in the remainder of this section. The third is discussed in section 4.1.

---

[c] Up to this point in the chapter, we have talked in terms of processes being composed of *activities*. To discuss role-based workspaces, however, we use the notion of *tasks*. Tasks differ from activities in that: an activity is composed of tasks, a task is an assignable work package, and a task is (almost always) assigned to one role with the role having responsibility of assuring that the task is successfully carried out. We make this distinction because role-based workspaces are intended to help agents complete their assigned tasks in the context of activities carried out by some group of agents (i.e., some team).

A workspace is analogous to a desk in a physical working environment. On or in the desk will be the documents the agents are working on, the tools they use in doing and checking their work, devices for communicating with other agents, and items that help the agents organize their work. More specifically, a workspace corresponds to a desk dedicated to one of the agent's assignments, and an agent would have several desks, one for each of his/her assignments.

This indicates a primary benefit of workspaces: the elimination of context-switching overhead. To switch contexts — to switch from working on one assignment to working on another — the agent moves from one desk to another; and when turning attention to an assignment, the agent finds the desk associated with this assignment the same as when he/she last left it. Further, because the desk is assignment-specific rather than agent-specific, much of the overhead associated with delegating or re-assigning work may be eliminated.

To better support today's rapidly growing information-intensive workforce, many of the items on/in a desk are typically moved to a computer. Workspaces can be viewed as an attempt to move as many items as possible (as well as reasonable) off the desk and into the computer. The resulting benefit is again a reduction of overhead effort. Providing electronic access, and allowing automated support, eliminates many assignment-specific overhead activities (for example, searching through a pile or file of documents) and greatly simplifies others (for example, finding the tools needed to work on a document).

3.2.2 Role-based Workspace Organization

A role-based workspace may be either *specific* or *generic*. Both focus on a particular project and a specific role for that project. A specific workspace additionally focuses on a specific task (or small subset of highly coupled tasks) that agents must carry out when occupying the role. Generic workspaces, on the other hand, reflect information pertinent to all the tasks in a project that are pertinent to the role.[d]

An agent accesses a specific workspace by selecting a task (or a strongly coupled subset of tasks). This creates a more narrowly scoped workspace reflecting just this task (or subset of tasks). The agent uses this specific workspace to carry out the work needed to perform the task(s).

An agent opens a generic workspace by first selecting a project to work on and then selecting a role within that project to occupy. The projects that the agent may select are constrained by the allocation of the agent to projects. The roles that may be selected reflect not only resource allocation decisions by the project's manager but also any role-occupancy constraints specified in the role's definition. As an example of the latter, an agent may not open a generic workspace that requires knowledge of a specific analysis tool if the agent's description does not indicate that the agent has this knowledge.

When a workspace is opened, information relevant to the specific agent, the chosen role and the chosen project is assembled and displayed in an organized way. The displayed workspace reflects all the aspects of the process relevant to the role. The contents and organization of the ge-

---

[d] Workspaces provide access to task-, role- and project-specific information. When working on a task, an agent additionally needs access to relevant personal and corporate-wide information. Relevant personal information includes the agent's schedule and to-be-done list, information that is typically stored in an agent's PDA. Relevant corporate-wide information includes: contact information for other personnel, information about training courses for various tools and techniques, and the identification of personnel who have volunteered to mentor with respect to various tasks or tools. Here, we assume that access to this information is provided outside of a workspace (for example, in an organization's corporate-wide knowledge base).

neric workspace reflect information about the agent's abilities, skills and preferences.

3.2.3 Specific Workspaces

Specific workspaces provide the basic, fundamental support for working on assignments. In the simplest situation, only one agent is working on the task. Frequently, however, two-or-three agents may collaboratively work on the task. In this section, we discuss the simplest, single-agent, situation. The more complex, multi-agent, situation is discussed in section 4.1.

Fig. 3 depicts an example of a specific, single-agent workspace.

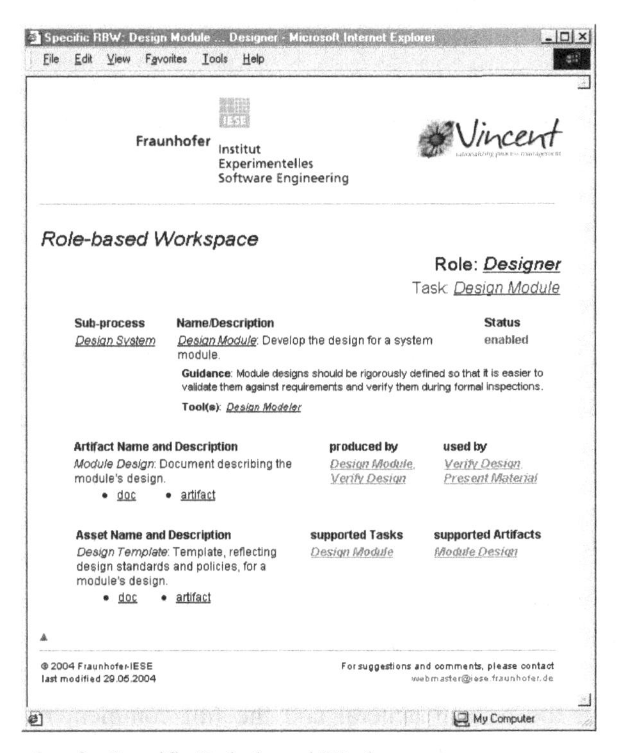

Fig. 3. Example of a Specific Role-based Workspace.

This example illustrates the basic items that constitute a specific workspace:

- Task Descriptions: Descriptions of the steps involved in carrying out the task as well as inter-dependencies among the steps. This provides the agent with the basic information about what has to be done when carrying out the task.
- Advice: Guidance concerning how to carry out the task successfully, including lessons learned and the experiences of agents who previously occupied the role. This allows the agent to benefit from previous experience in carrying out the task.
- Artifacts: Descriptions of the artifacts that the role works on when carrying out the task as well as links to the actual artifacts needed when or resulting from performing the task. This information allows the agent to understand the task, and review the result of performing it, in terms of its effect on the artifacts used and produced during task performance.
- Tools: Descriptions of the tools the agent needs in performing the task as well as links to the tools' documentation and the tools themselves. This information allows the agent to efficiently access the tools and consult tool documentation. It can also guide the agent in choosing among a collection of alternative tools.
- Assets: Descriptions of relevant templates, checklists, etc., as well as links by which to download copies of these assets or view policy, guideline, standards and reference documents. This information provides the agent with a task-specific "doorway" into the usually very large collection of assets the organization has accumulated over time.

3.2.4 Generic Workspaces

A specific workspace provides the basic support an agent needs to rationally carry out a specific task. A generic workspace, on the other hand, is role-specific but does not pertain to any specific task. It reflects information about the role in general and the full complement of tasks in which the role participates. Its major intent is to help agents occupying the role to properly achieve the role's purpose in a process and properly function as a member of the project team performing the process.

Fig. 4 and Fig. 5 illustrate a generic workspace. These figures show that, to assist agents, a generic workspace includes information about:

- Responsibilities: Descriptions of the role's overall obligations and permissions with, where possible, links to the tasks, artifacts, and other process elements pertinent to meeting the obligations within the constraints levied by the permissions. This helps the agent understand and focus on his/her responsibilities.
- Task List: A list of the tasks for which the role is responsible. As new tasks are assigned they are added to this list. The list provides the role with a continuously up-to-date agenda for his/her work.
- Process Documentation: A structured collection of links into relevant parts of the process handbook. This allows the agent to efficiently consult the process description as needed in the course of his/her work.

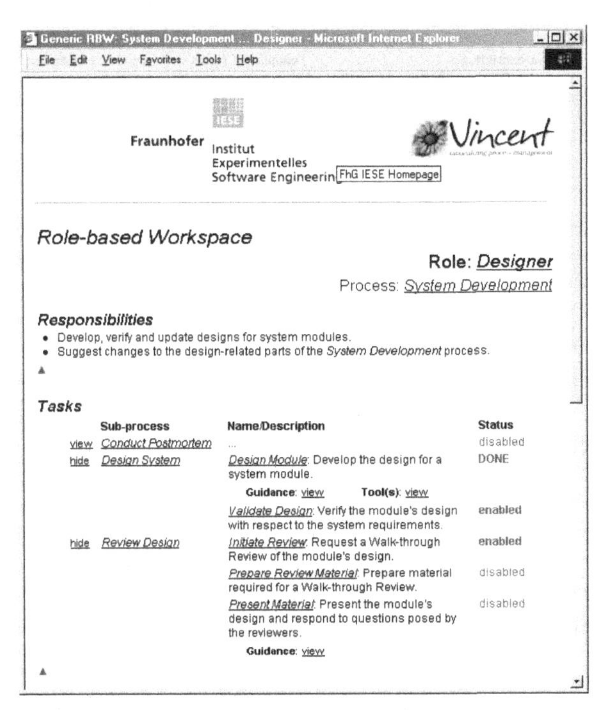

Fig. 4. Example of a Generic Role-based Workspace (Responsibilities and a selection of assigned Tasks).

This information not only helps agents plan and track their work, but also helps them better understand the constraints upon their work and the rationale for this work as part of the overall process being performed.

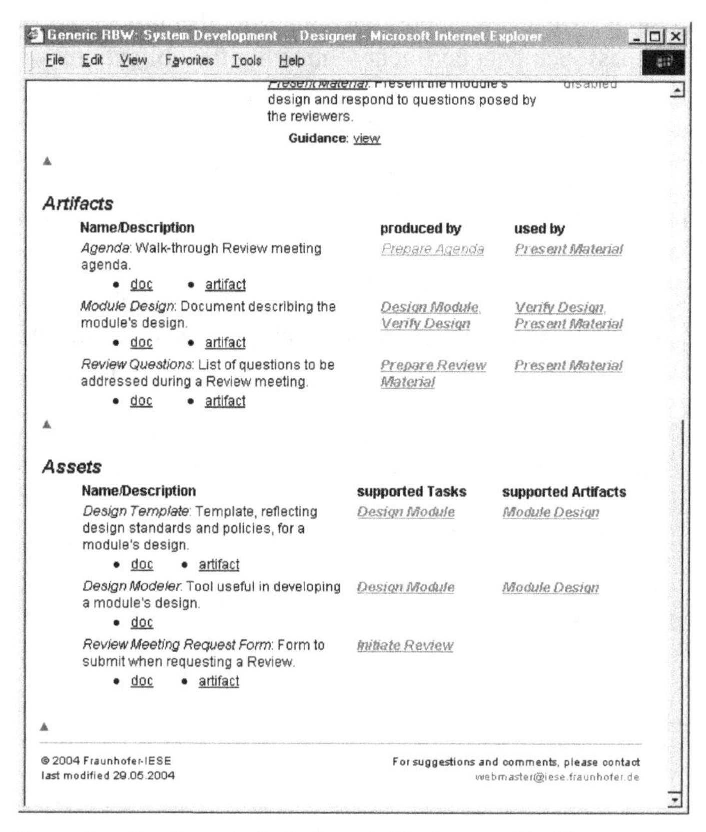

Fig. 5. Example of a Generic Role-based Workspace (Artifacts and Assets related to viewed Tasks).

3.3 Software Project Control Centers

The complexity of software development projects is continuously increasing. This results from the ever-increasing complexity of functional as well as non-functional software requirements (e.g., reliability or time constraints for safety critical systems). The more complex the requirements are, the more people are usually involved in meeting them, which only further increases the complexity of controlling and coordinating the project. This, in turn, makes it even harder to develop the system according to the plan (i.e., matching time and budget constraints). Coordination issues are usually addressed by Computer Supported Cooperative Work (CSCW) systems which ease communication among project members and support document sharing. CSCW support is further discussed in section 4.1. The focus in this section is on support for controlling a project.

Project control issues are very hard to handle. Many software development organizations still lack support for obtaining intellectual control over their software development projects and for determining the performance of their development processes and the quality of the produced products. Systematic support for detecting and reacting to critical project states in order to achieve planned goals is usually missing[24].

One way to support effective control of software development projects is the use of basic engineering principles[25, 26], with particular attention to the monitoring and analysis of actual product and process states, the comparison of actual with planned states, and the initiation of any necessary corrective actions during project execution. Effectively applying these principles requires experience-based project planning[27]; that is, the capture of experience from previous projects (such as activities, measurement plans, and baselines), and the use of explicitly defined models reflecting this experience, in order to plan a project. Furthermore, it requires the collection, interpretation, and presentation of measurement data according to a measurement plan; that is, the establishment of measurement-based feedback mechanisms in order to provide stakeholders with up-to-date information about the project state. Moreover, it requires experience packaging after project completion so that future projects are influenced by the experience gained in previously-performed projects.

In the aeronautical domain, *air traffic control systems* are used to ensure the safe operation of commercial and private aircraft. Air traffic controllers use these systems to coordinate the safe and efficient movement of air traffic (e.g., to make certain that planes stay a safe distance apart or to minimize delays). The system collects and visualizes all critical data (e.g., the distance between two planes, the planned arrival and departure times) in order to support decisions by air traffic controllers.

Software project control requires an analogous approach that is tailored to the specifics of the process being used (for example, its non-deterministic, concurrent, and distributed nature).

A *Software Project Control Center* (SPCC)[24] is a control system for software development which collects all relevant data to project control, interprets and analyzes the data according to the project's control needs, visualizes the data for different project roles, and suggests corrective actions in the case of plan deviations. An SPCC could also support packaging of data (e.g., as predictive models) for future use and contribute to an improvement cycle spanning a series of projects.

Before discussing existing SPCC approaches, we first discuss the notion of "controlling a project". *Controlling a project* means ensuring the satisfaction of project objectives by monitoring and measuring progress regularly in order to identify variances from plan during project execution so that corrective action can be taken when necessary[28]. Planning is the basis for project control and defines expectations which can be checked during project execution. The gathered experience can be packaged for future projects after project completion in order to support organization-wide improvement cycles. All corrective actions needed to bring a project back to plan – that is, all *steering activities* – are explicitly included in the notion of "controlling a project".

A Software Project Control Center is a means for interpretation and visualization of measurement data during process performance and therefore supports controlling a project. An SPCC has a logical architecture that clearly defines interfaces to its environment, especially to all project members relying on SPCC information, and a set of underlying techniques and methods that support controlling a project.

From a more technical perspective, an SPCC utilizes data from the current project (e.g., the project's goals, characteristics, baselines, and measurement data) and experiences from previous projects (e.g., information captured in quality, product, and process models) and produces a visualization of measurement data by using the incorporated techniques and methods to interpret the data. An SPCC is a general approach to project control and is not necessarily tool-supported. But in order to successfully, efficiently carry out control activities such as monitoring defect profiles, detecting abnormal effort deviations, cost estimation, and root-cause analyses of plan deviations, a certain amount of tool support is necessary and inevitable.

In the following we highlight two SPCC approaches addressing different objectives. The first deals with integrated approaches; that is, SPCC approaches that are tightly integrated into the project's performance and act as a focal point for all project issues and organizational improvement efforts. The second deals with goal-oriented data processing and visualization; that is, presenting data regarding different project needs and supporting different project stakeholders.

3.3.1 Integrated Controlling Approaches

Integrated SPCC approaches are tightly integrated into every project's performance and are actively used to gain experience for future projects. Such approaches are normally used by organizations to improve the maturity of their software processes and practices and establish organization-wide standards.

One example is NASA's *Software Management Environment (SME)*[29, 30], which was developed by the Software Engineering Laboratory (SEL)[31, 32] at the NASA Goddard Space Flight Center (GSFC). The main aim of this SPCC is to support the manager of a software development project by providing access to three information sources: (1) The *SEL Database* holding information from previous projects; that is, subjective and objective process and product data, plans, and tool usages. (2) The *SEL Research Results* database holding different models (such as growth or effort models) and relationships between certain parameters/attributes (described with quality models). Primarily, this information may be used to predict and assess attributes. (3) The *SEL Management Experience* database holding information about project management experiences in the form of rules within an expert system. The rules help inexperienced managers analyze data and guide re-planning activities. For example, this database includes lists of errors and appropriate corrective actions.

All this information is input for an SME, which uses it to perform management-oriented analyses fostering well-founded decision-making. Experience gained during project execution may lead to changes of the information. This feedback mechanism enables the SME to work with up-to-date information.

3.3.2 Goal-oriented Data Visualization

A Goal-oriented SPCC approach (G-SPCC) is a state-of-the-art framework for project control developed at the University of Kaiserslautern and the Fraunhofer Institute for Experimental Software Engineering (IESE)[33, 34]. The aim of this approach is to present the collected data in a goal-oriented way in order to optimize a measurement program and effectively detect plan deviations.

The purpose of the G-SPCC approach is to support agents occupying roles. Project control is driven by different role-oriented needs. We define *control needs* as a set of role-dependent requirements for obtaining project control. A project manager needs different kinds of data, data of different granularity, or different data presentation modes than a quality assurance manager or a developer. For example, a manager is interested in an overview of the project effort in order to compare it to previously defined baselines, while a developer is interested in the effort she/he spent on a certain activity. As another example, a quality assurance manager is interested in the efficiency of a certain inspection technique, while the project manager is primarily interested in a list of defects and how many defects have to be fixed in order to release a product to the next project phase. In general, control-oriented information needs differ between more management-oriented project roles (such as a project manager or a quality assurance manager) and more technically oriented roles (such as a tester or a programmer). The first group is more interested in charts presenting an overview of the overall project, while the second is more interested in activities within the project. However, it is important to note that control-oriented information needs may vary, significantly, within these groups.

Fig. 6 gives an overview of the G-SPCC architecture. It shows that measurement data is collected during project performance and interpreted with respect to the goals and characteristics of the project as well as project plan information (e.g., baselines, number of project members, and developer skills) and control needs (e.g., the kind of control technique that should be applied, and tolerance ranges). The outputs of this interpretation (performed by *SPCC functions*) are displayed by a set of *SPCC views*, each providing role-specific insights into the process (e.g., insights suitable for project managers, quality assurance personnel, or the development group). The SPCC interpretation and visualization process is supported by an experience base in order to reflect data from previous projects and store experience gathered after project completion.

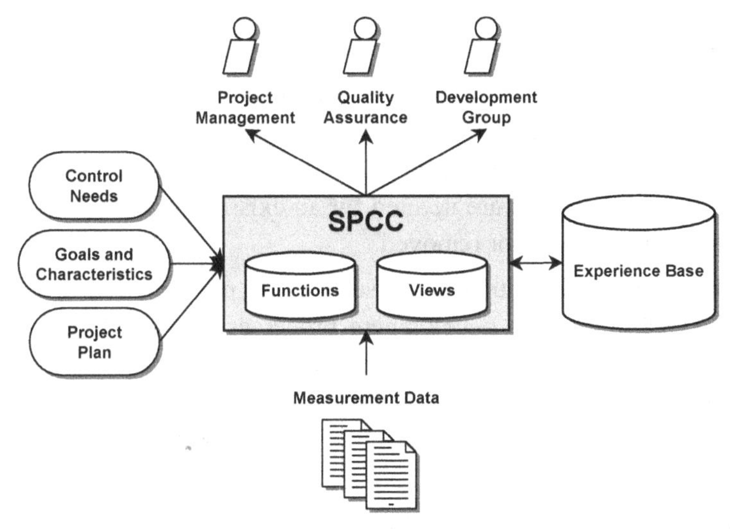

Fig. 6. The G-SPCC Architecture to Support Different Project Roles.

The G-SPCC approach is based on the Quality Improvement Paradigm (QIP)[35] and consists of the following steps:

- First, project stakeholder control needs are characterized in order to set up a measurement program able to provide a basis for satisfying all needs.
- Then, measurement goals are defined and metrics are derived determining what kind of data to collect. The GQM paradigm is used to derive these metrics and create a set of data collection sheets that are assigned to certain process steps. The process is modeled using SPEARMINT®/EPG (described in section 3.1).
- Next, a *Visualization Catena* (VC) is defined to provide online feedback on the basis of the collected data. The VC includes a set of con-

trol techniques and methods corresponding to the measurement goals. For example, a suitable Tolerance Range Checking technique may be included to detect baseline deviations. A VC also includes a set of views to visualize project data.

- Once the VC is specified, a set of role-oriented views are defined to support control of the project. As measurement data are collected, the VC analyzes and visualizes them accordingly. For example, an SPCC function detects baseline deviations and a corresponding view displays them.
- Once a deviation is detected, its root cause must be determined and the VC adapted accordingly. A baseline deviation, for example, can lead to new or adapted measurement goals and baselines. In this case, new VC components are defined for an existing VC or existing components are changed or removed.
- After project completion, the resulting VC may be used to improve future VCs, software development processes, and baselines. For example, views of the generalized effort progression can be used to improve the baselining of future projects or process efficiency views may be used to enhance the definition of process steps.

The benefits of the G-SPCC approach include: (1) improvement of quality assurance and project control by providing a set of custom-made views of measurement data, (2) support of project management through early detection of plan deviations and proactive intervention, (3) support of distributed software development by establishing a single point of control, (4) enhanced understanding of software processes, and improvement of these processes, via measurement-based feedback, and (5) preventing information overload through custom-made views with different levels of abstraction.

4 Using Process Information

The purpose of this section is to illustrate how displayed process information may be used to support process performers in doing their work. First, the usage of role-based workspaces (introduced in section 3.2) is discussed, including their customization and personalization as well as coordination and collaboration issues. Second, some sample control techniques based on a Software Project Control Center (introduced in section 3.3) are discussed. Subsequent to this, two sections briefly address more advanced concepts, namely the improvement-oriented organization of experience (including process information) and support for process enactment (that is, providing more proactive support for enacting a defined process).

4.1 Using Role-based Workspaces

Previously, we discussed role-based workspaces as a way to collect and organize the information pertinent to an agent when occupying a role and performing some task. In this section, we first discuss the tailoring of role-based workspaces to meet the needs and capabilities of specific role occupants. We then discuss the various ways in which role-based workspaces may provide significant support for coordination and collaboration, helping an agent perform more effectively, efficiently and accurately as a member of a project team.

4.1.1 Customization

Different agents will have different capabilities stemming from their differing levels of education, training and experience. Effective role-based support requires that role-based workspaces may be customized to match a role's specific capabilities. Customization involves:

- Varying the content: Inexperienced, novice agents need more support than experienced, expert ones. For example, guidance might be included in a novice's role-based workspace but not included in a role-based workspace for an expert. Also included in a novice's, but not an expert's, workspace might be: detailed rather than synoptic descriptions of the tasks; explanations of required skills; an indication of the factors affecting successful task completion; as well as other information.
- Providing access to available support: Novices will also benefit from being able to easily access support for their work. This could include

an indication of professional development courses related to a role's required skills. It could also include an identification of role-related mentors accompanied by their contact information.

- Varying the display of information: Agents will vary with respect to their basic approach to solving problems. An often-noted, and very major, difference is that between graphically-oriented problem-solvers versus textually-oriented ones. This implies that different modes should be available for displaying process-related information. For example, an activity decomposition structure might be displayed as hierarchically indented text or graphically as a tree.
- Varying the organization of information: Another often noted, and also major, difference among agents is whether they approach problem-solving in a bottom-up (inductive) or top-down (deductive) manner. This also implies that different information-display modes should be available. For example, drop-down lists might be available to allow the agent to navigate through the information in a top-down (deductive) manner.

Obviously, information about an agent's capabilities is needed in order to provide customized role-based workspaces. Much of this information can come from providing agent profiles identifying expected collections of capabilities and having the agent identify the most appropriate profile when establishing a role-based workspace. Alternatively, the program which establishes a role-based workspace could hold a pre-programmed dialog with the agent to obtain information affecting the workspace's content, organization, and rendition.

4.1.2 Personalization

Agent capability differences will also lead to different patterns of role-based workspace usage. In addition, agents will accumulate – over time – their personal arsenal of resources they have found useful in carrying out their work. Finally, agents will gather observations and advice about how to effectively, efficiently, and accurately carry out the work.

Therefore, effective role-based support requires that agents be able to personalize their role-based workspaces. Personalization involves:

- Re-defining the content: It is impossible to predict all the ways in which role-based workspaces should be customized to meet individual agent needs. It should be possible for agents to delete information from and add information to the workspace.
- Re-organizing the display: It is also impossible to predict all the ways individual agents will want to position the information displayed in a workspace. It should be possible for agents to combine, split and re-position the information within the role-based workspace's display.
- Adding assets: Agents will, over time, accumulate many personal assets they have found valuable in performing their work. It should be possible for agents to add these assets to their workspaces.
- Annotating the displayed information: In analogy to writing marginal notes in a paper-based process handbook, it should be possible for agents to add notes to pieces of information in their workspaces.

Supporting personalization along these lines requires a variety of workspace support capabilities. Many of these capabilities imply the inclusion of a variety of ways to display information on desktops and within windows; providing these capabilities depends on the capabilities provided by the support system's underlying operating-system contexts (e.g., MS 2000 vs. MS 2003). Others concern the manipulation of information provided by web page displays. Regardless of how they may be achieved, the capabilities include:

- Tracking of desktop and window contents and organization accompanied by reconstitution of the configuration when the desktop or window is re-opened.
- Drag-and-drop manipulation of information displayed on a desktop and within windows (for example, repositioning text or other elements within the display).
- Drag-and-drop manipulation of information displayed within web page displays (for example, repositioning the items in some graphic included in a web page).
- Annotations of documents and web-pages (for example, providing text input boxes within a web page).

- Dynamic changes to information display modes (for example, changing from a drop-down list to a tiled set of windows).

All of the cited example capabilities are provided by current, state-of-the-practice technology. In fact, current technology might support even more extensive capabilities and even more effective workspace personalization. Empirical studies of the possibilities, and their importance, is needed to determine what can be provided and its value.

4.1.3 Coordination

A major benefit of role-based workspaces is that they can support the coordination of work either by an agent over time or among a group of agents carrying out inter-related tasks. By coordination we mean the effective, efficient, accurate performance of tasks in an orderly fashion as affected by inter-task constraints. We discuss this benefit in this section.

A very simple case of coordination concerns the ability of an agent to easily switch among the various roles he/she may occupy. An agent may have several generic and specific workspaces, perhaps from different projects, open at some point in time. However, only one specific workspace and its relevant generic workspace will be *active*. This helps agents work on assignments in a well-focused way. It also helps them rapidly and efficiently switch among their various assignments. Current operating-system contexts fully support this *context switching*.

More significant is the support role-based workspaces can provide for organizing and guiding an agent's work within the context of a process. A particularly useful, yet quite simple, situation is helping the agent focus only on those tasks that he/she may work on. As indicated previously, the generic workspace includes a list of the tasks the role is responsible for or participates in. This list may indicate which tasks are *enabled* — may be worked on — and which tasks are *blocked* — may not be worked on because some pre-condition is not satisfied. This allows the agent to quickly focus his/her attention. (Fig. 3 provides an example of displaying information about the *enabled* status of tasks.)

Additional support along these lines may focus on other aspects of carrying out the tasks. This support relates to the status of the tasks and can include:

- Condition Checkers: These are tools the agent may use to check the completeness, correctness and accuracy of his/her work. Process descriptions will often identify, usually in the terms of the status of artifacts, conditions that must be achieved, for example, that a `Design` activity must result in a `well-organized Design` artifact. To the extent that the `Design` being `well-organized` can be checked by analyzing the `Design`, the agent can receive assistance in checking this condition.
- Events: Often, conditions may not be checked in an automated way. For example, the condition that a `Design` document is `correct` is usually checked by a combination of desk-checking and design reviews. To support task enabling and blocking, the fact that the `Design` document is `correct` needs to be recorded. A role-based workspace should provide the ability to record the satisfaction of conditions so that this information may be used to coordinate the agent's future work.
- Task Dependencies: It is often the case that the tasks being performed by one agent are enabled by the work completed by other agents – this is, in fact, the most usual situation. Therefore events, in general, lead to the enabling of tasks in a workspace. This leads to changes in the enabled/blocked status of the tasks in some other workspace. Active "announcement" of status changes – for example, by some visual or aural signal such as those typically used to announce "you have mail" – will help the agent keep up with changes to his/her work.

In summary, one major purpose of coordination capabilities is to allow agents to assess the completeness and accuracy of their work.

Another major purpose is to coordinate work across several tasks being carried out by several agents. This requires capabilities that allow agents to signal each other about the status of their work. This could lead to quite extensive coordination support. For example, if a task is blocked, the agent might use the Process Documentation items in a generic workspace to track back through pre-condition/post-conditions to locate predecessor tasks and then use a Task Status Query to get information about the source of the blockage. If necessary, the agent may then interact with other agents (using capabilities discussed in the following section) to collaboratively understand and solve problems inhibiting progress. This allows the agents to collectively work in a very focused way, making their collective work more effective and efficient.

There is an additional possibility for supporting cross-agent coordination. In the task list, additional information may be displayed about enabled tasks to indicate tasks for which the satisfaction of one or more pre-conditions has been re-established since the agent previously completed the task. This flags tasks the agent might have to re-do because some process element, probably an artifact, has been changed. The agent may open specific workspaces for these flagged tasks, use task dependency information in the workspaces or process handbooks to understand what has changed, and carry out the task again as needed.

4.1.4 Collaboration

Inter-agent coordination support is particularly valuable when the agents are geographically distributed. Event-based enabling of items on task lists provides significant coordination support in this case. But much more is needed – the agents additionally need support for the collaboration that, were they geographically co-located, would be accomplished by organized or informal face-to-face meetings and discussions of their work and any problems which arise.

Two possibilities arise when considering collaboration within a group of geographically separated agents. One case – *synchronous collaboration* – occurs when the agents can all work at the same time. In the other case – *asynchronous collaboration* – the agents must for some reason (availability, time zone differences, etc.) work at different times. We discuss the use of role-based workspaces to support synchronous and asynchronous collaboration in the remainder of this section.

One of the major goals of work in the field of Computer Supported Cooperative Work (CSCW) has been to support geographically distributed teams working synchronously. (For a general introduction to CSCW work, see the overview by Grundin et al.[36].) CSCW work suggests that this support includes at least the following basic capabilities:

- Agenda: A meeting agenda provides both a plan for a specific real-time meeting and a place to record decisions and action items.

- Audio: Support for communication by voice.
- Shared Windows: Display of information at multiple various workstations with changes made by the "owner" of the displayed information propagated to the displays.
- Shared Whiteboard: A window displayed on all the workstations that all participants can modify using drawing/text capabilities they would typically use in writing on a whiteboard in a meeting room.
- Real-time Chat: Broadcasted and directed, person-to-person, transmission of commentary to allow participants to record their thoughts and share them with others.

Additional capabilities may be included to support specific needs. For example, a video image capture/display capability might be added to allow broadcasting of a (physical) whiteboard in one of the agents' offices.

These basic capabilities <u>could</u> be added to a specific workspace to support synchronous collaboration. CSCW work to date, however, provides a much simpler solution. This work has led to many distributed meeting systems, several of which are commercially available. Because these focused and integrative solutions exist, generic workspaces do not have to be extended to provide the capabilities. Rather, they may be used in tandem with these other solutions.

Representative commercial products are Microsoft's *NetMeeting*[37] and Teamware's *Pl@za*[38]. Another example is the eWorkshop system developed at the Fraunhofer Institute at the University of Maryland[39]. These distributed meeting systems may be used to provide the needed agenda, audio, shared whiteboard and real-time chat support. An example collaboration support window, resulting from using the eWorkshop system, is shown in Fig. 7.

Collaboration across tasks additionally requires substantial support for coordinating the mutual influences and constraints among the tasks as specified in the process definition. For example, the process description might indicate the flow of artifacts among tasks by indicating how the tasks produce and consume the documents. As another example, the process definition might specify, or at least imply, precedence relations among the tasks.

Collaboration across tasks may be supported by the following basic capabilities:

- Asynchronous Communication: Mail-style interactions among the agents.
- Shared Document Spaces: A shared file structure that all of the agents can access and modify.
- Threaded Discussions: A means to raise questions about, add support to, and refute comments about some issue as well as spawn new issues.

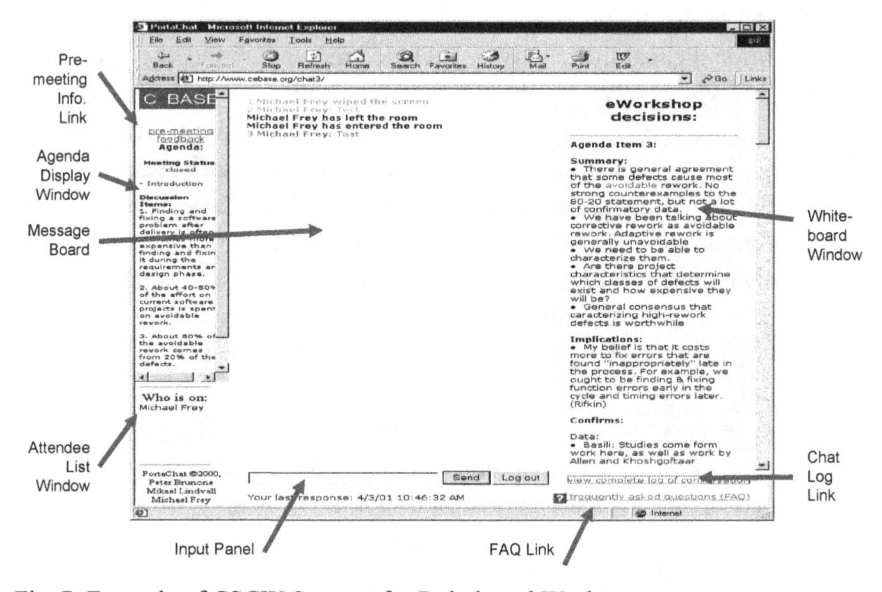

Fig. 7. Example of CSCW Support for Role-based Workspaces.

E-mail is the normal approach to supporting asynchronous communication. An example system developed to provide shared document spaces is the Basic Support for Collaborative Work (BSCW) system[40]. Threaded discussions about documents are also supported by BCSW. Support for threaded discussions in general has affected the support provided by most currently available commercial systems. CSCW work has led to systems — Teamware's *Pl@za* is, again, an example — that integrate support for asynchronous communication, shared document spaces, and threaded discussions into systems that support distributed meetings.

4.2 Using Software Project Control Centers

Section 3.3 introduced Software Project Control Centers (SPCCs) as a primary means to support project control and discussed how to provide a role-oriented visualization for gathered measurement data. The focus of this section is on how an SPCC may be utilized by a specific project role; that is, what kind of concrete support can be provided. Before we give examples of this support, we first address the basic purpose of project control; i.e., we address the purposes of an SPCC application.

The following discussion uses concepts adapted from the Software Management Environment (SME) approach[29]:

- *Monitoring*[e] refers to observing a project's state and progress by observing attributes, or combinations of attributes, of the project's processes, products, and resources.
- *Comparison* uses archived data from completed projects or nominal performance guidelines to judge the progress and health of the current project.
- *Analysis* focuses on (1) examining the monitoring results, and (2) applying information about a project's context to identify the probable causes of deviations from the nominal performance guidelines.
- *Assessment* analyzes, with weighting, information about the project to form a judgment of project, product, and process quality.
- *Prediction* extrapolates information about attributes of a project's processes, products, and resources from the current project status to assess the future project behavior. In general, prediction always requires some kind of mathematical model. Fenton and Pfleeger[41] define prediction as identifying relationships between various process and product factors and using these relationships to predict relevant external attributes of products and processes.

[e] In the SME approach this is called *observation*.

	Module A	Module B	Module C	Module D	Total
Planned Value (PV)	20	40	30	10	100
Earned Value (EV)	20	30	10	0	60
Actual Cost (AC)	18	32	17	0	57
Schedule Variance = EV - PV	0	-10	-20	-10	-40
Cost Variance = EV - AC	2	-2	-7	0	-7

Table 1. Earned Value Sample for Module Costs in Thousand US Dollars.

- *Planning* defines baselines or a nominal value for certain measures. In addition, it focuses on assessing (alternative) planning decisions and their effects. This is the basis for further dynamic re-planning during the execution of the project.
- *Guidance* proposes a number of courses of action according to a specific situation or an identified problem. Based on these proposals, a manager might be able to initiate corrective actions and take steps to improve or enhance the development process. A developer, on the other hand, might use guidance as assistance for harmonizing his/her own performance with respect to the overall process and given project goals.

The core of an SPCC is the set of integrated project-control techniques and methods. These methods usually cover different purposes, such as monitoring project attributes, comparing attributes to baselines, or predicting the course of an attribute. The most advanced support includes assessing the overall state of the project and guiding a project participant through corrective actions if a project differs from its plan. Assessing a project's overall state may be achieved by *Earned Value Approaches*[42], which identify important key factors for assessing the overall project state. Project participant guidance may be achieved by building upon experience from previous projects.

4.2.1 Assigning Overall Project State

The Earned-Value Approach is a management technique used to assess the current state of a project. It was first defined at the end of 19th century when engineers decided to determine cost efficiency by comparing the actual cost (AC) of work performed with the earned value (EV) and the planned value (PV).

Table 1 illustrates the basics of this approach: It concerns the control of the costs (measured in thousand US dollars) of creating four software modules, A to D. For each module, we have planned values derived during the planning phase and the actual costs determined by data collection procedures. As work is performed, the earned value of this work is measured on the same basis as for the planned values. Several techniques may be used to compute the earned value for a certain work package; these are beyond the scope of this chapter. However, to provide a simple example: if a work package is complete, its earned value is equal to the planned value, whereas, if a work package has not started yet, its earned value is zero. For example, in Table 1, module A is complete, work on module D has not yet started, and one-third of module C's planned value has already been achieved.

When controlling a project, project managers are interested in the *schedule variance* (that is, the earned value minus the planned value) and the *cost variance* (that is, the earned value minus the actual cost). In the example shown in Table 1, and with respect to the point in the project reflected in the table, work for 40,000 US dollars remains to be done, but 7,000 US dollars have been expended without any recognition of its value. A simplistic conclusion is that the project plan has been violated. An earned-value conclusion, however, would compare the actual cost with the planned values and not indicate a plan violation.

4.2.2 Experience-based Approach

The first technique we address in this context is the *Sprint I* approach[43, 44, 45], built upon clustering algorithms to dynamically adapt the prediction of key project attributes during project execution. Sprint I is not a pure approach to project control according to our definition because it predicts project attributes before the project starts and, therefore, covers planning as well as performance concerns. However, the Sprint I approach provides an example of how a project manager might use experience from previous projects in order to control an on-going project. The prerequisite for a successful application of Sprint I is that a software development organization has performed a number of similar projects and measured at least one key attribute (e.g., effort per development phase) for each of these projects. Additionally, there is the requirement that the context for each of these projects (i.e., the boundary conditions such as organizational, personal and technical constraints) has been characterized.

Briefly, the technique is as follows: First, context-specific measurement data from former projects is analyzed in order to identify clusters. Based on the context of the project to be controlled, the technique selects a suitable cluster and uses the cluster curve (the mean of all curves within a cluster) for predicting the attributes to be controlled. During the performance of the project, the prediction is modified based on actual project data. This leads to an empirically based prediction and, as a result, to flexibility for project and context changes.

The second experience-based approach to applying SPCCs was developed in the context of NASA's SME. It capitalizes on experience gained in previously-performed projects. Doerflinger and Basili[46] describe the use of dynamic variables as a means to monitor software development and provide guidance in case of plan deviations. The core notion is to assume underlying relationships that are invariant across similar projects. These relationships are used to predict the behavior of projects with similar characteristics. A *baseline* for a certain project attribute (such as the effort in person-hours) is derived from measurement data for one or more completed projects within the same context. The baseline is used to determine whether the project is in trouble or not. If the current values of a variable fall outside a *tolerance range* (i.e., a predetermined tolerable variation from the baseline), the project manager is alerted and has to determine the possible reasons for the failure. A number of tables list possible reasons for deviations above or below the tolerance range for each measured project attribute. If a particular reason appears more often than some other reason, the former is assumed to be more probable than the latter.

4.3 Process Enactment Support

So far, we have focused on enabling, facilitating the work of agents in carrying out their assigned tasks. The capabilities we have discussed serve to present information about what should happen, what has happened and the status of a project in qualitative (status related) and quantitative (measurement based) terms. Agents interpret this information to focus on the tasks they need to do, understand how they can carry out these tasks, and gain access to supportive resources.

More proactive support is possible. With this proactive support, there is some control over focusing the agents' attention, directing his/her work, and promoting the use of specific resources. In general terms, with proactive support: (1) an automated system makes decisions rather than allowing the agents to make decisions, and (2) the automated system implements some of the actions that the agent might carry out.

A simple form of proactive support is *workflow management*. In this case, the flow of artifacts, as defined in the process handbook, is used to actively focus the attention of agents on enabled tasks and actively manage the flow of the actual artifacts as agents complete their tasks. Another, more extensive, form of proactive support is *enactment support*. In this case, the status of, and measurements about, all process elements, not just artifacts, is used to focus agent attention and manage the process performance.

Numerous workflow management and enactment support systems have been developed and they have focused on many software development process issues. The majority of them have been designed to support the performance of a project by tracking states and actively guiding developers based on this state information. Examples range from flexible workflow management systems, to object-oriented database systems, to distributed, open environments. The systems developed to date have been decidedly immature because of the complexity of the goal. Nevertheless, some success stories do exist, such as LEU[47] from LION GmbH, Germany, and ProcessWEAVER[48] from Cap Gemini, France.

4.4 Experience Management

This section discusses the ways in which experience from former projects may be reused for planning and controlling new projects. Basically, we can distinguish among organization-wide experience (such as general effort baselines for a certain project type, lessons learned from former projects, process models, product models, etc.) and project-specific experience (such as project plan information, instances of models, schedules, specific effort distributions, etc.). We define an *experience base* to be a repository of both organization-wide and project-specific experience described with respect to context-specificity and significance, for example, the experience pertains to a specific context (e.g., valid for all projects of a certain domain) and is stated with respect to a certain significance (e.g., a model has been successfully applied in five projects).

One approach to providing an experience base is an Experience Factory[35, 49]. This approach is based on the Quality Improvement Paradigm (QIP) approach to evolutionary process improvement. The QIP approach comprises the following basic steps: (1) Characterize, (2) Set Goals, (3) Choose Process, (4) Execute, (5) Analyze, and (6) Package. Each of the six steps can be interpreted in both a project-specific and organization-wide manner:

- The steps can be characterized as follows for a specific software development project. (1) Characterize the project environment; that is, determine the project type and elements to be reused from an experience base. (2) Set quantifiable goals; that is, define quality goals and select corresponding models, specify hypotheses, and identify influencing factors for the hypotheses. (3) Choose the right process and define a project plan; that is, specify how the defined goals should be achieved. (4) Execute the project according to the previously defined plan; that is, perform the planned development activities, manage the project, and collect measurement data. (5) Analyze project results; that is, compare hypotheses with real data and identify deviations and their reasons. (6) Package project experience; that is, capture project information in the project-specific experience base and update the organization-wide experience base (e.g., add a new effort baseline for a specific project context, update the significance of an existing model, or correct relevant models).
- The steps can be interpreted as follows for the whole organization: (1) Characterize the organization and identify trends. (2) Define general improvement goals and corresponding quantifiable hypotheses. (3) Choose pilot projects for validating the goals and hypotheses. (4)

Execute the pilot projects, collecting data regarding the goals and hypotheses. (5) Analyze the results and, in particular, validate the hypotheses. (6) Package project experience in the form of reusable experience elements and update/refine the existing experience base.

Fig. 8 illustrates the packaging step following project completion and focused on reusing and updating an effort model. The experience base, at the top of the figure, indicates the original state before packaging experience from a specific project, P. Model M1 has been used to forecast the effort distribution for this project. Model M1 is valid for projects of context C1 and has already been applied in S former projects. During analysis and packaging, the project results may lead to three different cases:

- Case 1: Model M1 was correct for project P; that is, the real effort distribution of project P was consistent with the distribution of model M1. In this case, the significance of model M1 is increased by one; that is, M1 has now been successfully applied in S + 1 projects.

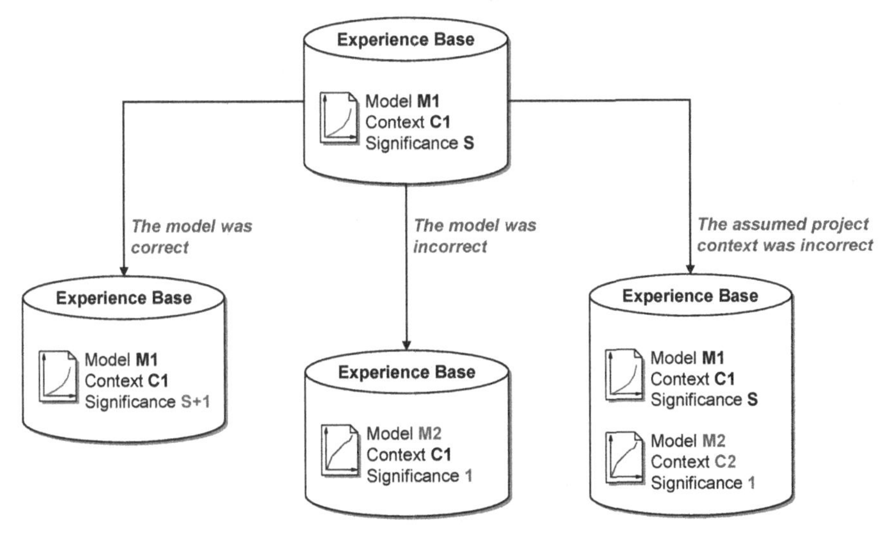

Fig. 8. Example for Packaging Project Experience.

- Case 2: Model M1 was incorrect for project P; that is, the real effort distribution of project P differed significantly from the effort distribution of model M1 (and no abnormal circumstances were recognized for project P). In this case, the original model has to be changed in order to reflect the new project experience. Model M1 is replaced by model M2 for context C1 and the significance of M2 is set to one, because M2 has been applied in only project.

- Case 3: The assumed context for project P was wrong; that is, a certain environment was expected (e.g., developer experience = `high`), but an analysis has shown that this original assumption was incorrect (e.g., the real developer experience = `low`). Therefore, the project context is actually a new context, C2. In this case, a new model M2 is added for context C2 with significance one and the original model is left unchanged for context C1.

5 Summary

This chapter describes how to capture, display, and use process information to support people performing a process. In addition, it provides many examples of specific capabilities that have proven valuable in practice.

First of all, we identified the needs for supporting people-oriented software development, discussed the relationships among roles, agents, and human process performers, and identified a variety of human characteristics and their influence on software engineering.

Then we provided an overview of different kinds of process information and how to collect this information prior to, during, and following process performance. We discussed the value of general, qualitative, status information about a process. In addition, we described one way to provide methodical, goal-oriented, support for collecting quantitative measurement data.

After that, we discussed a variety of gradually more significant ways in which people may be supported in carrying out their assigned tasks:

- First, we described simple support provided by process handbooks describing an agents' responsibilities, the activities they perform, the artifacts used and produced by the activities, resources supporting activity performance, and the conditions that reflect progress.
- After this, we discussed more extensive support provided by creating role-based workspaces that collect together all the information an

agent needs to access when carrying out a task or a set of inter-related tasks.

- Then we discussed even more extensive support that can be provided when the information displayed to agents not only reflects the process definition but also reflects, qualitatively and quantitatively, what has happened during process performance.

Subsequently, we presented extensions to this basic support which facilitates coordination and collaboration among agents cooperatively carrying out their tasks.

Finally, we discussed the opportunity to use the status information and measurement data to actively, automatically control process performance. In addition, we addressed how to use status information and measurement data not only to proactively support process performance (through enactment support) but also manage the experience gained during project performance.

Acknowledgments

Proof-of-concept work on specific and generic role-based workspaces was done by Mirjam Krementz[50]. The early work on EPGs was, in addition to the authors of the cited reference[16], influenced by Alan Christie, Denis Avrilionis and Anthony Earl. The PMC process documentation generation capability was, in addition to the authors of the cited reference[18], influenced by Dave Barstow, Bill Cohagan, Michael Mahan, Don Oxley and Dick Phillips. The following people participated in the development of SPEARMINT®/EPG: Danilo Assmann, Ulrike Becker-Kornstaedt, Fabio Bella, Dirk Hamann, Ralf Kempkens, Holger Neu, Alexis Ocampo, Peter Rösch, Louise Scott, Martin Soto, Martin Verlage, Richard Webby, and Jörg Zettel. The following students contributed to the implementation of SPEARMINT®/EPG: Martin Denn, Ralf Hettesheimer, Thomas Kiesgen, Dietmar Klein, Arne Könnecker, Dirk Ludwig, Sascha Schwarz, Slavomir Sobanja, Martin Vogt, and Christian Voigtländer.

References

1. B. Curtis, M.I. Kellner, and J. Over. *Process modeling*. Communications of the ACM, 35(9), pp. 75-90 (1992).
2. H.D. Rombach and M. Verlage. *Directions in software process research*. In Marvin V. Zelkowitz, editor, Advances in Computers, vol. 41, pp. 1-63. Academic Press (1995).
3. A. Endres and D. Rombach. *A Handbook of Software and Systems Engineering – Empirical Observations, Laws, and Theories*. Pearson Education Limited, Addison Wesley (2003).
4. IEEE 1058.1 — *Software Project Management Plans*. IEEE Computer Society Press, Los Alamitos, California (1987).
5. M. Paulk, B. Curtis, M. Chrissis, and C. Weber. *Capability Maturity Model for Software (V1.1)* (CMU/SEI-93-TR-024, ADA 263403). Software Engineering Institute, Carnegie Mellon University, Pittsburgh, Pennsylvania, USA (1993).
6. Pragma Systems Corporation, Reston, Virginia 20190, USA. [*http://www.pragmasystems.com*]
7. Integrated Process Asset Library, Federal Aviation Administration (FAA), Washington, D.C., USA. [*http://www.faa.gov/ipg/pimat/ipal*]
8. W. Scacchi. *Process Models in Software Engineering*. In: J. Marciniak (ed.), Encyclopedia of Software Engineering, 2nd Edition, Wiley (2001).
9. U. Becker, D. Hamann, and M. Verlage. *Descriptive Modeling of Software Processes*. In Proceedings of the Third Conference on Software Process Improvement (SPI '97), Barcelona, Spain (1997).
10. M. Verlage. *About views for modeling software processes in a role-specific manner*. In Proceedings of the second international software architecture workshop (ISAW-2) and international workshop on multiple perspectives in software development (Viewpoints '96) on SIGSOFT '96 workshops, San Francisco, California, United States, pp. 280-284 (1996).
11. R.W. Selby. *Amadeus Measurement-Driven Analysis and Feedback System*. In Proceedings of the DARPA Software Technology Conference, Los Angeles, CA (1992).
12. R. van Solingen and E. Berghout. *The Goal/Question/Metric Method - A Practical Guide for Quality Improvement of Software Development*. McGraw-Hill International (UK) (1999).
13. V.R. Basili, D.M. Weiss. *A Methodology for Collecting Valid Software Engineering Data*. In IEEE Transactions on Software Engineering, Vol. SE-10, No. 6, pp. 728-738 (1984).
14. V.R. Basili, G. Caldiera, and H.D. Rombach. *Goal Question Metric Paradigm*. In: "Encyclopedia of Software Engineering", Volume 1, edited by John J. Marciniak, John Wiley & Sons, pp. 528-532 (1994).

15. L.C. Briand, C.M. Differding, and H.D. Rombach. *Practical guidelines for measurement-based process improvement*. Software Process: Improvement and Practice, 2 (4) (1997).
16. M. Kellner, U. Becker-Kornstaedt, W. Riddle, J. Tomal, M. Verlage. *Process Guides: Effective Guidance for Process Participants*. Proceedings of the Fifth International Conference on the Software Process: Computer Supported Organizational Work, Chicago, Illinois, pp. 11-25 (1998).
17. U. Becker-Kornstaedt, J. Münch, H. Neu, A. Ocampo, and J. Zettel. *SPEARMINT® 6. User Manual*. Fraunhofer Institute for Experimental Software Engineering, Report 031.02/E, Kaiserslautern (2003).
18. W. Riddle and H. Schneider. *Coping with Process Agility*. Tutorial at 2002 Software Engineering Process Group Conference (SEPG 2002), Phoenix, Arizona (2002).
19. W. Riddle. *Coping with Process Specification*. In Proceedings of Integrated Design and Process Technology (IDPT-2003) Conference, Austin, Texas (2003).
20. S. Miller and W. Riddle. *Experience Defining the Performance Management Key Process Area of the People CMM ... and the Link to Business Strategy*. In Proceedings of 2002 Software Engineering Process Group Conference (SEPG 2000), Seattle, Washington, (2000).
21. A. Ocampo, D. Boggio, J. Münch, G. Palladino. *Towards a Reference Process for Wireless Internet Services*. IEEE Transactions on Software Engineering, Special Issue on Wireless Internet Software Engineering, 29 (12), pp. 1122-1134 (2003).
22. F. Bella, J. Münch, A. Ocampo. *Capturing Experience from Wireless Internet Services Development*. In Proceedings of the International Conference on Applications and Emerging Trends in Software Engineering Practice (STEP 2003), Workshop on "Where's the evidence? The role of empirical practices in Software Engineering", Amsterdam, The Netherlands, pp. 33-39 (2003).
23. J. Zettel, F. Maurer, J. Münch, L. Wong. *LIPE: A Lightweight Process for E-Business Startup Companies Based on Extreme Programming*. Lecture Notes in Computer Science 2188, (Frank Bomarius and Seija Komi-Sirviö, eds.), Springer-Verlag, pp. 255-270 (2001).
24. J. Münch and J. Heidrich. *Software Project Control Centers: Concepts and Approaches*. Journal of Systems and Software, 70 (1), pp. 3-19 (2003).
25. W.W. Gibbs. *Software's Chronic Crisis*. Scientific American, pp. 86-95 (1994).
26. M. Shaw. *Prospects for an Engineering Discipline of Software*. IEEE Software 7(6), pp. 15-24 (1990).
27. H.D. Rombach and M. Verlage. *Directions in Software Process Research*. Advances in Computers 41, pp. 1-63 (1995).

28. Project Management Institute. *A Guide to the Project Management Body of Knowledge (PMBOK® Guide) 2000 Edition.* Project Management Institute, Four Campus Boulevard, Newtown Square, PA 19073-3299 USA (2000).
29. R. Hendrick, D. Kistler, and J. Valett. *Software Management Environment (SME)— Concepts and Architecture (Revision 1)*; NASA Goddard Space Flight Center Code 551, Software Engineering Laboratory Series Report SEL-89-103, Greenbelt, MD, USA (1992).
30. R. Hendrick, D. Kistler, and J. Valett. *Software Management Environment (SME)—Components and Algorithms*; NASA Goddard Space Flight Center, Software Engineering Laboratory Series Report SEL-94-001, Greenbelt, MD, USA (1994).
31. L. Landis, F. McGarry, S. Waligora, R. Pajerski, M. Stark, R. Kester, T. McDermott, and J. Miller. *Managers Handbook for Software Development—Revision 1*; NASA Goddard Space Flight Center Code 552, Software Engineering Laboratory Series Report SEL-84-101, Greenbelt, MD, USA (1990).
32. F. McGarry, R. Pajerski, G. Page, S. Waligora, V.R. Basili, and M.V. Zelkowitz. *An Overview of the Software Engineering Laboratory*; Software Engineering Laboratory Series Report SEL-94-005, Greenbelt, MD, USA (1994).
33. J. Heidrich and M. Soto. *Using Measurement Data for Project Control.* In Proceedings of the Second International Symposium on Empirical Software Engineering (Vol. II), Rome, pp. 9-10 (2003).
34. J. Heidrich. *Effective Data Interpretation and Presentation in Software Projects.* Technical Report 05/2003, Sonderforschungsbereich 501, University of Kaiserslautern (2003).
35. V.R. Basili, G. Caldiera, and H.D. Rombach. *The Experience Factory.* Encyclopedia of Software Engineering 1, pp. 469-476 (1994).
36. J. Grundin, S. Poltrock, and J. Patterson. *CSCW Overview.* Special Presentation at ACM Conference on Computer-Supported Cooperative Work (CSCW'96), Boston, USA (1996).
37. NetMeeting, Microsoft Corporation, Redmond, Washington 98052-6399 USA. [*http://www.microsoft.com/windows/netmeeting*]
38. Teamware Pl@za, Teamware Group Oy, Helsinki, Finland. [*http://www.teamware.net/Resource.phx/twplaza/index.htx*]
39. V. Basili et al. *Building an Experience Base for Software Engineering: A report on the first CeBASE eWorkshop.* In Proceedings of the 3rd International Conference on Product Focused Software Process Improvement (PROFES 2001), pp. 110-125, Kaiserslautern, Germany (2001).
40. W. Appelt. *WWW Based Collaboration with the BSCW System.* In Proceedings of SOFSEM'99, Springer Lecture Notes in Computer Science 1725, Milovy, Czech Republic, pp. 66-78 (1999). [*http://bscw.fit.fraunhofer.de*]

41. N.E. Fenton and S.L. Pfleeger. *Software Metrics. A Rigorous and Practical Approach.* International Thomson Computer: 2nd edition, London, UK (1996).
42. ACQWeb. Earned Value Management. [*http://www.acq.osd.mil/pm*]
43. J. Münch, J. Heidrich, and A. Daskowska. *A Practical Way to Use Clustering and Context Knowledge for Software Project Planning.* In Proceedings of the 15th International Conference on Software Engineering and Knowledge Engineering (SEKE 2003), pp. 377-384, San Francisco, USA (2003).
44. J. Münch and J. Heidrich. *Using Cluster Curves to Control Software Development Projects.* In Proceedings of the First International Symposium on Empirical Software Engineering (Vol. II), Nara, pp. 13-14 (2002).
45. J. Münch and J. Heidrich. *Context-driven Software Project Estimation.* In Proceedings of the Second International Symposium on Empirical Software Engineering (Vol. II), Rome, pp. 15-16 (2003).
46. C.W. Doerflinger and V.R. Basili. *Monitoring Software Development Through Dynamic Variables.* In Proceedings of IEEE Conference on Computer Software and Applications (COMPSAC), pp. 434-445 (1983).
47. G. Dinkhoff, V. Gruhn, A. Saalmann, M. Zielonka. *Business Process Modeling in the Workflow Management Environment LEU.* In Proceedings of the 13th International Conference on the Entity-Relationship Approach (Lecture Notes in Computer Science, Vol. 881, pp. 46-63). Berlin, Heidelberg, New York: Springer (1994).
48. Christer Fernström. *ProcessWEAVER: Adding process support to UNIX*. In Proceedings of the Second International Conference on the Software Process: Continuous Software Process Improvement, Berlin, Germany, pp. 12-26 (1993).
49. R. Basili. *Quantitative Evaluation of Software Engineering Methodology*. Proceedings of the First Pan Pacific Computer Conference, Melbourne, Australia (1985).
50. M. Krementz. *Personal Workspaces for Electronic Process Guide (EPG) Users*. Project Thesis, Fraunhofer Institute for Experimental Software Engineering, Kaiserslautern, Germany (1999).

Chapter 6

REQUIREMENTS AND VALIDATION OF THE E3 PROCESS MODELING SYSTEM

Letizia Jaccheri

Department of Computer and Information Science
Norwegian University of Science and Technology (NTNU)
Sem Sælands vei 7-9 7491 Trondheim, Norway
email: letizia@idi.ntnu.no

In the framework of the general goals of this book, which are to discuss the state of the art of software process topics and provide practitioners with a practical view of the developed methods, I use my own experience as a process modeling researcher over the last fifteen years to make my own contribution to the goals. I will address the following questions. How have process modeling researchers elicited requirements for software process modeling systems? First, to what extent have users been involved in the definition of these requirements? Second, how has IT evolution contributed to this requirements definition? Lastly, how has general software engineering knowledge influenced this definition? E3 (Environment for Experimenting and Evolving Software Process Models) is a process modeling system conceived to provide help for process/project managers, who construct and maintain models, and for practitioners, who use software process models. The initial requirements of the E3 system have been derived from a literature survey, lessons learned by working with other PM systems, and use of general-purpose technology for process modeling purposes. E3 has been designed and implemented twice. The first version of the E3 system was validated by a case study and the results of this validation resulted in the requirements for the second version of the system. The second version of the E3 system has been validated by empirical investigations in industrial and academic settings. The answers to the research questions given in this chapter have not to be regarded as an attempt to provide a general state of the art of software process topics or a complete view of the field for practitioners. Rather, they have to be considered as a set of lessons learnt about the development and validation of one of the process modeling systems that have been developed in recent years.

1. Introduction

The E3[a] project started in 1992. E3 offers a process modeling language and a supporting system for model construction, change, and inspection. A functioning prototype is available at [12]. In this work, I[b] look back and I try to reconstruct the research process that led to requirements definition, design, implementation, and validation of the system. The goal of this paper is to give an answer to one main research question.

How have process modeling researchers elicited requirements for software process modeling systems?

I investigate this question trying to explain the *why* dimension of software process modeling and to say something, looking at the past, about the future of process modeling systems. I further decompose the main question into three.

- RQ1: How much user involvement was there in the definition of these requirements?
- RQ2: How has IT evolution contributed to this requirements definition?
- RQ3: How has general software engineering knowledge influenced this definition?

When I think of E3, I first remember all the students and colleagues who have been involved in the project. And my memories are brightened or darkened by the joy of mutual understanding, the disappointment at discovering the wrong choices, a longing for the old times, the energy of starting a research project with almost no funding. I still experience sadness and anger for the times we were not understood by reviewers and funders. I remember curiosity, frustration, and the feeling of finally learning something new. And again I sense happiness and satisfaction at getting papers published at international conferences and journals and our work accepted by organizations. To organize my recall in a scientific way and not be overwhelmed by emotions, I use two sources of information: the E3 system and the articles which we have written about the system over the years. The research method which I use in this work is a critical review of our own work. On the other hand, the research methods, which have been exploited during

[a]E3: Environment for Experimenting and Evolving software processes, read E cube.

[b]Here, the form I is used when the text refers to the reconstruction of requirements. The form we is used when the text refers to choices made and activities done in the general context of the E3 project.

the life of the E3 project encompass both engineering and empirical based research methods and will be among the topics of this paper. This paper is structured as follows. Section 2 can be read in two ways. The fragments per se provide a snapshot of the story of the E3 project. The comments (or analysis) of each fragment gives a reflection about the fragment in the light of the three research questions. Section 3 provides a summary of the features of the E3 system version 2 as is available at [12]. Section 4 provides further discussion of the relationships between requirements and the research questions in this paper. Section 5 concludes the paper by giving indications for further work.

2. On the nature of the E3 requirements

Here, I choose some fragments from the papers published about E3 to reconstruct the story about how the system was conceived and validated. I choose those fragments that give insight into the research questions of this work.

- RQ1: How much **user** involvement was there in the definition of these requirements?
- RQ2: How has **IT evolution** contributed to this requirements definition?
- RQ3: How has general **software engineering knowledge** influenced this definition?

By users, in this chapter, I mean all actors who have something to do with a process model. This definition of user encompasses both process designers, process owners, and process performers.

The research methods were either empirical (when we performed some kind of empirical investigation), engineering based (when we reported about software design issues), or theory based (when we made choices made as to software engineering theories).

Section 2 2 is further divided into three. Section 2.1 is about the first activity in the E3 project, when we used Coad and Yourdon OO analysis methods and languages for software process modelling. Section 2.2 and section 2.3 are about the first and the second versions of the system and their validation.

This section can be regarded as a kind of post mortem[6] analysis of the E3 project. Next section 4 provides a further analysis of the fragments in the E3 context.

2.1. *Use of existing OO analysis methods and languages for process modeling*

Fragment 1:

The PM community has produced many PM systems [21 1 8 53] *that use and experiment with various process modeling paradigms. The experience with using these systems is still limited in modeling both processes that are standardized by organizations, e.g. ISO, DoD and processes that are adopted by large software factories. ⋯ The principal goal of E3 is to get hands-on experience with using an object-oriented paradigm on real-world software processes. We will put emphasis on experiments with modeling software processes, rather than the goal of finding a new PM paradigm or language.* (From [15])

The process is based on literature survey and knowledge about other PM systems. Fragment 1 introduces an issue about the nature of the modeled processes. Processes standardized by organizations, like ISO and DoD, are written process descriptions, which are not always consistent with the actual processes. At the same time, large organizations usually have written process descriptions. What is not written here is what other research projects had not modeled, like for example, processes in small organizations, which are not necessarily formalized by quality manuals or standardized. In this way, this fragment is about the influence that real users (organizations and standards) have had on the E3 research process. In fragment 1 there are references to five papers [21 1 8 5 3].

This is not only a literature review but also of a living research network that was active in the late 80s and early 90s in the PM field. That network was mainly European and initiated the European Workshop of Software Process Technology and the European Promoter project. I had been working on both the EPOS [14] and Oikos [1] project and had knowledge of the design choices and features of the two environments. From this perspective, this text fragment binds the E3 research process to software engineering community knowledge.

Fragment 1 is about experimentation. Given that related research projects have devoted a lot of efforts to the development of new languages and execution engines, the initial choice was not to spend resources on the implementation of yet another process modeling language. On the contrary, we decided to reuse existing technology for experimentation. The other systems had been validated against standards and organization processes. We did not declare which kind of processes we wanted to address. Experimentation opened up interaction with users. The experimentation choice was dictated by the trends in software engineering knowledge at the time (early 1990s).

Fragment 2:

The E3 process modeling framework will offer graphical data-model to design process models as sub-models of activities, software products, tools, development organizations and resources. In addition a process model should capture constraints like temporal aspects, control flow among activities and product, and sequencing among activities, resource allocation, connections to tools, responsibility, and communication. (From [16])

Fragment 2 is about the nature of process models. The decisions of choosing activities, software products, tools, development organizations and resources as building blocks of a process model and constraints like temporal aspects, control flow among activities and product, and sequencing among activities, resource allocation, connections to tools, responsibility, and communication derived from related software engineering theories.

Fragment 3:

We will experiment with an object relation model [20] *that will allow us to structure models through aggregation and relations. Objects, types, and relations must be explicitly represented and persistent. Relations will be used to express constraints, also at the type level. We want to exploit relations as much as possible to make our process models more declarative and to make explicit the dependency among the different components.* (From [16])

Fragment 3 is about the decision to exploite object orientation. The choice was influenced by both software engineering theories and the availability of OO technology on the market.

Fragment 4:

We exploit as much as possible existing technology, commercially available software packages. We foresee the integration of a DBMS offering both object relation datamodel and concurrent transactions, C++ for activities programming, and a user interface system. There is already a number of C++ based frameworks and libraries available for user interface construction, inter-process communication, persistency, and database management. (From [16])

Fragment 4 is about the IT to be used for the development of a PM system. Fragment 4 implicitly makes choices about the software architecture. The architecture will rely on a DBMS with concurrent transactions. It is not specified if concurrency will be allowed among modelers or among performers. We declared that we wanted to use C++ for activity programming. This sentence means that we were planning to provide execution support, and activities to be executed had to be programmed in C++. Here, we must recall that other PMLs were based on programming languages at the same abstraction level as C++. SPELL [14] was based on Prolog and Arcadia [21] was based on the Ada language.

Fragment 5:

A software process model will change over time The modeling framework must also assist in the process of changing process models, and cope with the effects of the changes. (From [16])

Fragment 5 opens up an important process modeling topic at the beginning of the 1990s, namely process evolution. This research was supported by theoretical work [18]. Other PM systems, like EPOS [14] and Spade [4] had paid considerable attention to evolution.

Fragment 6:

Our first experience in modeling a process with OO techniques used the Coad and Yourdon [7] OO analysis and design methods and supporting tools to model the process of a department of the FIAT car manufacturer. (From [16])

Fragment 6 is a general declaration of intent about the process to be followed and its relation to users (a department of the FIAT car manufacturer). Specific OO techniques from software engineering will be exploited. IT is mentioned, there is a reference to the intention of exploiting supporting tools, but the specific tools are not mentioned here. We decided to model the Iveco quality manual as for some practical reasons that manual was available to us since we had some personal contacts at that organization. We decided to use OO theories. The rationale for this choice was that thedy were the mainstream theories for software design and programming at the time and that these theories were supported by available languages and tools.

Fragment 6 is about the choice of the first validating users (a department of the FIAT car manufacturer) and the kind of process to be modelled (a quality manual). Moreover, there is a reference to the choice of exploiting OO in general and Coad and Yourdon specifically.

Fragment 6 is in contrast with fragment 1. While in 1, we criticize the choice of other PM research projects of modeling only quality manuals of big organizations, here we choose to do the same. It is much easier to model a process manual than eliciting the process of an organization as it is. The latter activity requires a lot of insights in one organization and mutual trust between the organization and the modeling team, which we did not have.

Fragment 7:

In order to provide a simulation tool for our experiments, we also developed a prototype PSEE based on distributed OO programming techniques. The experience with the PSEE was resisted by the organization management that hardly perceived it as a real asset and considered it inapplicable on an organization-wide scale. Based on this experience, we focused our research on the elicitation of process models through object orientation. (From [16])

Fragment 7 is about an important choice in E3: the choice of not including the execution (or enaction) requirement. The choice comes from interaction with users.

Fragment 8:

This preliminary experience in using OO design and analysis techniques used on an as-is basis for process modeling demonstrated that it is possible to model a software process at a high abstraction level by using pure OO analysis techniques without delving into low-level details. Moreover, our experience also demonstrated the effectiveness of object orientation in eliciting process models, since the models were used also as a means to communicate information to the process users. (From [16])

Fragment 8 is about the validation of the first E3 experience. We were able to communicate the OO models back to the users who owned the process manuals, i.e., the quality manager and his group. This fragment does not describe what users liked or did not like about the offered features.

Fragment 9:

Nevertheless, despite the encouraging results, OO design and analysis techniques used on an as-is basis revealed some problems ... (From [16])

Fragment 9 is an introduction to fragments 10 and 11.

Fragment 10:

Syntax and semantics are not defined formally, thus preventing automatic analysis or simulation. (From [16])

Fragment 10 is about the drawbacks of applying OO analysis methods directly to PM. At the present time of writing there are OO analysis methods with formally defined syntax and semantics. We regard fragment 10 as an IT issue. Since we did not have available languages (and supporting tools) with formal syntax and semantics, we set this as a requirement for a new language. Fragment 10 and all the fragments about language requirements and above all the choice of implementing the E3 PML are in contrast with fragment 1. After declaring that E3 should not develop a new language but rather invest effort in investigations, we decided to implement a new language only after one modeling attempt with one modeling language and tool. When I look back at that choice it seems to me that we did not reflect enough on our initial choices and intentions or on this decision to implement a new PML.

Fragment 11:

Process-dedicated syntax constructs are needed in order to enhance process understanding. Although our experience showed that the techniques we employed increase process understanding, they also indicated that nontrivial process models can consist of hundreds of classes and associations that appear to the user as a flat web of identical boxes and arrows. Hence, in order to enhance understanding, process-specific constructs mapped on the process components are needed, still using a graphical notation. (From [16])

Fragment 11 derives from general software engineering knowledge. The idea is the same as that of predefined types in programming languages and it is an instantiation of the reuse theory. In E3 we combined this idea with that of assigning a special graphical syntax to the predefined classes and associations.

Fragment 12:

The Iveco model encompasses 161 classes and 585 associations. Since it does not make sense to present more than circa 10 classes in a single page, one needs policies to section the model for presentation purposes. When inspecting an OO process model, the data and control flow perspective are of primary importance. In our context, data flow means that for a given task class, one is interested in seeing which are the input and output classes, etc. In addition to the classical data and control flow, for a given task, it is useful to find out which are its responsible agents, and which tools it uses. (From [13])

Fragment 12 is about the view mechanisms which have been found useful during the validation with users.

Fragment 13:

The Smalltalk simulation showed absence of trivial errors, e.g., deadlocks. However, it cannot be regarded as a true simulation in which probabilistic parameters are assigned to activities and resources as was suggested by the process owners. Also, the manual translation from E3 PML to Smalltalk can introduce errors. We have then abandoned this research path and we have decided to focus on static analysis instead. Static analysis is more suitable than dynamic simulation if the purpose of the models is understanding by humans and not execution. This assumption is supported by the fact that it was difficult for the users to understand and appreciate the Smalltalk simulation. (From [13])

Fragment 13 is about the evaluation of the enaction feature and gives reasons why we abandoned this research path as a consequence of user interaction. Fragment 13 is consistent with fragment 7.

2.2. *E3 version 1 and its validation*

Fragment 14:

E3 PML enables class and association creation and definition. In the following, we will always refer to classes and associations and not to their instances. This is because the goal of our work is to provide descriptions of process manuals by means of process model templates. (From [2])

Fragment 14 is also about instantiation. Here, we declared that the goal is to describe process manuals (and not real world processes as declared in 1), but not to provide enactment. This choice was inspired by related work.

Fragment 15:

The E3 PML is an object-oriented language augmented with association management. It offers a set of kernel classes and associations with process modeling semantics. Kernel classes are organized in an inheritance hierarchy as described in the following. A class inherits all the attributes, comments, and methods of its super-class, and all the associations defined for its super-class. Inherited features can be re-defined in the sub-class. (From [2])

Fragment 15 is about the main characteristics of E3 PML version 1, which is an object-oriented language augmented with association management. This choice is driven by general software engineering knowledge.

Fragment 16:

Then, the tool offers four kinds of views, Inheritance, Task, Task Synchronization, and User, that implement respectively the OO (Inheritance view), functional (Task View), structural (Task Synchronization View), and informational (User View) perspectives (From [2])

The choice of adding different views comes from software engineering theories and methods such as data flow diagrams and control flow diagrams. For the first version of the E3 system, these views were conceived for both change (or editing) and inspection.

Fragment 17:

The main weaknesses of E3v1 (E3 version 1) revealed by this case study are: lack of instance level facilities, lack of a flexible view mechanism, problems in the execution support (From [16])

Fragment 17 is about validation of the first version of the E3 system in the context of the Iveco case study. The modeling requirements came from the users, i.e., the Iveco process owners and the students who modeled the process. The requirements derived from this validation were that there was the need to increase flexibility. The first version of the system enabled its users to navigate in a model of starting tasks. For a given task, it was possible to view its sibling tasks, its member tasks, and its related products, tools, and roles. In addition, the user view could display other items and their relationships in an unconstrained way. User Views were difficult to manage and not easy to use as they lacked a theory model. Flexible views needed to be defined. Also, the views provided by E3 p-draw v1 are task-oriented and do not enable the user to browse a model from a perspective that is different from the task view. For instance, it can be useful, for a given product to see which tasks consume it, or produce it, etc. This also applies to tools and roles.

Fragment 18:

E3 p-draw v1 does not support the Instance level. While a template is an abstract description for a set of models, a model is a description of a single process, including time and resource binding. If a template has to be understood and used, it must be possible to generate (either automatically or manually) instantiated models. (From [16])

From conversations with the users we found out that we were not able to communicate to them the advantages of the Smalltalk simulation. The users were looking for a representation of the resource and time plan. From these interactions with our users, we derived this requirement about instantiated models.

2.3. *E3 version 2 and its validation*

Fragment 19:

A related issue was represented by portability. Version 1.0 was developed in C++ under DEC Ultrix because of its reliance on the object-oriented DBMS Object/DB we used as a model repository, and the Interviews library was used for GUI programming. Nevertheless, it became clear that support for PC boxes was highly desirable, due to their increasing pervasiveness. The implementation of version 2.0 minimized portability concerns through the adoption of the Java language [19]. Java enabled portability on all platforms supported, which presently include PCs as well as Unix boxes, and provided a uniform API for GUI programming. (From [16])

Portability becomes an issue when moving E3 from a student context at university to industrial settings [10]. The Olivetti case study was performed by master students who had not participated in the requirement definition of the E3 system. These students interacted with a quality manager from Olivetti and his group. However, the case study had been designed in a way that the objective was more that of asserting the validity of the E3 features than getting contructive feedback from users.

On the other hand, at university, the execution environment may coincide with the development one. The portability discussion was crucial to the decision of adopting Java for the second implementation of E3. The implementation of E3 version 2 started a few months after the release of the language. Fragment 19 about the implementation language is about IT and portability.

Fragment 20:

Moreover, the use of a true object-oriented language opened up interesting developments as far as simulation is concerned. Currently, E3p-draw elements are mapped onto Java classes and objects. E3 presently leaves the behavior of methods unspecified. Specifying such behavior with the Java language would lead to a nice integration of E3p-draw with subsystems providing dynamic analysis, simulation, or even enaction. (From [16])

Fragment 20 is also about IT and opens up the question of enactment. Here it is interesting that the adoption of a new technology (Java) gave us some extra possibilities and we reconsidered the possibility of offering enactment.

Fragment 21:

In realistic process models like the one presented in the previous section, the number of process entities (e.g., tasks and artifacts) to be described tends to increase significantly. Consequently, developing a complete process model is a daunting activity that can seldom proceed in a straightforward top-down manner. In many situations, the only viable approach is to proceed both bottom-up and top-down until a reasonable description of the process is obtained. This requires flexible mechanisms to integrate multiple process fragments, which are often independently developed by different modelers. (From [10])

Fragment 21 is about the need to have modularization facilities. It comes from user interaction with special reference to the case study performed at Olivetti.

Fragment 22:

The rationale of introducing the check property presence and check property absence operations is that textual information can sometimes be preferable to graphic snapshots. For example, when E3 models have to be parsed and processed by other automatic tools. (From [10])

Fragment 22 is about the query mechanism. It is dictated mainly by compatibility issues with IT. This is also derived by the Olivetti case study.

Fragment 23:

The process modeling activities conducted during the past years have emphasized the importance of studying and understanding the associations among the entities of a process. The Olivetti experience confirmed this hypothesis. It is basically impossible to structure a process model statically in a way that any viewpoint or navigation path is smoothly supported. (From [10])

Fragment 23 is about the validation of the association concept in the Olivetti case study.

Fragment 24:

We report from an experiment in which we compared the E3 PML with respect to the standard modeling language IDEF0 for the purpose of model construction. The experiment has been run as part of a process improvement course in which forty students participated. Our hypothesis was that E3 will lead to less problems than IDEF0 when constructing software process models. (From [17])

Fragment 25:

As a conclusion from our data we are 90% sure that there will be less modeling problems when using E3 PML (From [17])

Fragments 24 and 25 are about one formal experiment for evaluation of E3v2 in academic settings. The experiment was run according to guidelines like those formalised in [22]. The hypothesis E3 will lead to less problems than IDEF0 when constructing software process models in the experiment context (a process improvement course taken by forty students). The objects of the experiments were E3 and IDEF0. The experiment was run at NTNU in 1999. The choice of evaluating the E3 system by a formal experiment is influenced by software engineering trends as the interest in formal experimentation was increasing in those years. We chose to run the experiment in a classroom setting as it would have been expensive to pay professionals to do the same modelling job as we asked the students to do. At the same time, I was teaching a course about software process improvement in which software process modelling was in fact a topic. As can be observed from this fragment, or more generally from the whole paper, the goal of the validtaion was not that of extracting requirements or getting directions for improvement.

Figure 1 shows an example of an E3 process model developed during the experiment reported in [17].

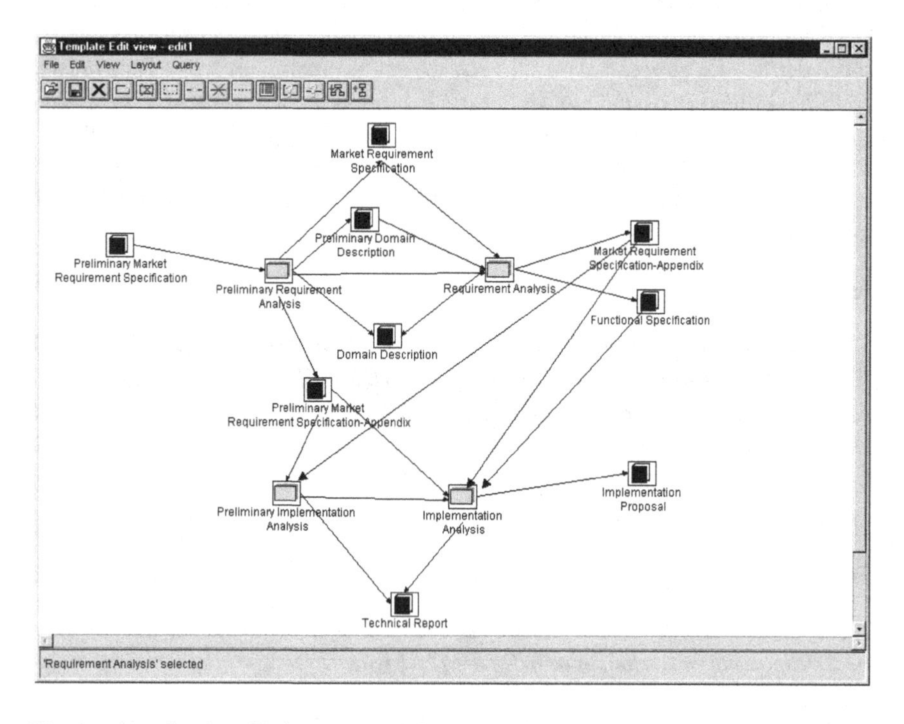

Fig. 1. An edit view displaying one of the five E3 models developed during the experiment.

Fragment 26:

The problem was the overview. Even with a rather simple process like this one it is difficult to maintain control. The fact that one has to model both horizontal and vertical relationships in addition to document flow contributes to this. (From [17])

In fragment 26 we report the two problems that students experienced when working with E3. Nevertheless the experiment reported in [17] was based on counting problems, here I interpret the reported problems.

This fragment is a negative validation of two E3 features: inspect by views (which should allow the user to keep control of the process model) and the kernel associations. The student here declares that it is difficult to keep control of a simple process model. He adds that horizontal and vertical relationships (we interpret these to be preorder and subtask) are difficult to combine with data flow (input and output).

3. E3 (version 2) PML and system: a summary

Here, there is a summary of the E3 features as they are provided by the existing implementation, that is, E3 version 2.

- E3 supports **modeling** of software development processes. Real-world processes can be represented by at least three kinds of process models:
  - Instance: Captures the full details of a project. Hence, it includes the mapping between the entities of a model and those of the real world. A model is, therefore, concerned with allocation of resources and deadlines because they are essential information for the project.
  - Template: Captures the key aspects of one or more Quality Manuals and Projects Manuals to describe the general issues that can be reused in the description of other similar processes, or to define a model which can provide guidance for a class of processes. In a template there is no concern about the mapping of model entities onto projects. A template can be refined into a new and more accurate template. This is the case for a Projects Manual being described as a refined and extended version of the template describing the corresponding Quality Manual.
  - Meta-level: the level at which the building blocks of a template are defined.
- E3 does not suppport process **enaction**.
- E3 supports **reuse** by inheritance, by instantiation from template to instance level, and by **module** facilities. Once a module is created, it encompasses the kernel, i.e. predefined classes and associations with the respective meta-classes.
- The **kernel** consists of **object-oriented classes** to describe tasks, products, roles, and tools and **associations** among these elements,

such as responsibility, preorder, aggregation, input, output, and connections between tools and tasks. If other kinds of entities need to be modeled, these must be created as specializations of the kernel classes. The same applies for associations.

- To create and **modify** new model elements, one needs to operate on edit **views**, which can be seen as workspaces.
- To **inspect** existing models, there are four kinds of derived views: simple, simple recursive, composite, and composite recursive. Basically, simple views visualize associations among classes, while composite views visualize aggregates together with the associations defined within them. A simple view is defined for a class and visualizes the class and all the association definitions the class participates in, except for aggregation associations. Derived views can be customized by hiding associations and classes. The user can specify which kind of associations need to be hidden and whether or not to visualize nodes connected by a currently invisible association. Each view (both workspaces and derived views) can be saved on persistent storage.
- Finally, base and derived views at the Template level provide an **automated instantiation feature**. The invocation of this operation generates a new base view at the Instance level containing an instance for every process element contained in the base view at the defined level. Additional instances can be defined by the user.
- E3p-draw provides a **query mechanism** to support static **analysis** of process models. The query mechanism will check the topology of the model, as determined by the definition of associations. For instance, it is possible to detect the presence of loops in an aggregation tree. More generally, E3p-draw provides support for checking whether or not a given property of association definition holds. For instance, one can check whether all the tasks of a given module have a responsible definition. Similarly, one can show tasks that lack a responsible definition. A query can be performed in the context of a whole module or of a view.
- E3p-draw is **portable** on all platforms which support Java.

4. Discussion

Table 1 shows the E3 requirements (rows) and the three dimentions given by the three research questions (columns). In the cells there are references to the fragments that give information about the relationships between requirements and research questions.

Each cell tells a short story about the given E3 feature and its origin (users, IT, or theories). When several fragments are associated with a cell, these can be inconsistent. This is because they may have been written in different periods and they may refer to different implementations of the system. Inconsistencies can be observed among fragments of the same row. I regard these inconsistencies as valuable as they tell about the system evolution and its causes.

Requirement	RQ1	RQ2	RQ3
modeling	1, 6, 24, 25	6	1, 2, 6
enaction	7, 13, 17	4, 20	
reuse by modularization	21		
OO classes associations	8, 12, 23, 26	3, 4, (9, 10) 19	3, (9,11), 15
inspection by views	17, 26		16
modification by views	17		5, 16
reuse by instantiation	17, 18		14
analysis by queries		22	
portability		19	

From the analysis performed in this section, we can generalize that when we say something about RQ1 (How much user involvement was there in the definition of these requirements?), we use fragments that have something to do with empirical investigations.

For example, looking at the intersection between *modelling* and *RQ1*, we find fragments 1, 6, 24, and 25. While 1 is about limitations in other PM systems evaluations, 6 is about the validation settings of the first industrial trial of the E3 system (at a department of FIAT) and Coad and Yourdon OO analysis and design methods and supporting tools. Fragments 24 and 25 are about a formal experiment run in academic settings to evaluate E3 version 2.

Column RQ2 (in Table 1) is about How has IT evolution contributed to these requirements definition?. If we read the fragments associated to this column, they are in general about engineering and design choices. Fragment 6 appears both in column RQ1 and RQ2 as it declares both the industrial validation context and the tools used for the trial. Fragment 4 is clearly about technology and its implication for enaction and object oriented modelling.

Column RQ3 (table 1) is about How has general software engineering knowledge influenced this definition? And here we generally refer to literature-based pieces of research. Fragment 1 is in both column RQ1 and RQ3 as it is about validation of other PM systems as discussed in the literature.

Another way to look at Table 1 is to read, for each requirement, associated fragments in ascending order. For example, if one takes into consideration requirement automated instantiation, fragment 14 says something about the initial choice, dictated by related work, of not including the instance level. Furthermore, fragments 17 and 18 tell about the validation of the first version of the system, the lack of instance level facilities and the rationale for introducing them.

The two requirements process modelling and OO classes and associations are those for which there are most associated fragments. This is somewhat natural, since E3 was conceived as a process modelling system based on OO augmented with associations. One could argue that OO classes and associations are not a requirement but a design choice. Another discussion topic is whether it is meaningful to assess these two requirements in their entireness or if it would have been more valuable to decompose them into smaller entities, like for example, to regard classes and associations as two distinct entities to be evaluated.

5. Conclusions and further work

The E3 project started in 1992. The experiment reported in [17] was run in 1999. In this chapter I have provided a summary of the features of the E3 system, a short story (given by article fragments) of the process that has led to the definition and implementation of E3, and a critical reflection about the definition, implementation, and validation of the system reqirements.

Many requirements are common nowadays. At the time of writing, it is common for software organizations to use electronic process guides supported by web-based intranet systems. On the other hand, requirements like the use of meta-level facilities to reuse process model knowledge are not commonly accepted.

This chapter is a contribution for those that want to learn about an existing PM system. E3 is also available at [12] and can be easily installed and tried.

The story of the E3 project can be used to plan future work with the system. We can choose which of the three research questions we want to address further. On the IT axis, one can look for which off-the-shelf components (both commercial and open source) can be exploited to re-engineer the system. On the software engineering axis, one can look at new theories, for example, in the area of component-based software engineering, measurement, and global software development. On the users dimension, there is a trade-off between elicitation of new system requirements versus validation of existing requirements.

This chapter makes clear that we have used a combination of research methods, empirical engineering, and theory. In this way, this chapter is a lesson learnt about experience with the different methods. As future work, we want to continue exploiting empirical-based research methods for eliciting requirements from users. At the same time, if we want to let our system evolve, we must work as engineers to incorporate new technology into the system. One idea is to make the E3 project into an open source project.

Acknowledgments

I thank all the students who have worked at the E3 project. Special thanks and thoughts go to Silvano Gai, Mario Baldi, Patricia Lago, Gianpietro Picco, Alfonso Fuggetta, Alessandro Bonaudo, and Marco Torchiano. Thanks to Reidar Conradi and Alf Inge Wang for comments and discussions about this chapter.

References

1. Vincenzo Ambriola, Paolo Ciancarini, and Carlo Montangero. Software Process Enactment in Oikos. In *Proc. 4th ACM SIGSOFT Symposium on Software Development Environments, Irvine, California*, pages 183–192, 1990.
2. Mario Baldi and Letizia Jaccheri. An exercise in modeling a real software process. In *AICA Italian Annual Conference*, 1995.
3. Sergio Bandinelli, Alfonso Fuggetta, and Carlo Ghezzi. Software Process as Real-Time Systems: A Case Study Using High Level Petri Nets. In [9], 1991.
4. Sergio Bandinelli, Alfonso Fuggetta, and Carlo Ghezzi. Software Process Model Evolution in the SPADE Environment. *IEEE Trans. on Software Engineering*, pages 1128–1144, December 1993. (special issue on Process Model Evolution).
5. Noureddine Belkhatir, Jacky Estublier, and Walcelio L. Melo. ADELE2 - An Approach to Software Development Coordination. In [9], pages 89–100, 1991.
6. Andreas Birk, Torgeir Dingsøyr, and Tor Stålhane. Postmortem: Never leave a project without it. *IEEE Software*, 19(3):43–45, 2002.

7. Peter Coad and Edward Yourdon. *Object-Oriented Analysis*. Prentice Hall, Englewood Cliffs, first edition, 1990.
8. Reidar Conradi, Espen Osjord, Per H. Westby, and Chunnian Liu. Initial Software Process Management in EPOS. *Software Engineering Journal (Special Issue on Software process and its support)*, 6(5):275–284, September 1991.
9. Alfonso Fuggetta, Reidar Conradi, and Vincenzo Ambriola, editors. *Proceedings of the First European Workshop on Process Modeling (EWPM'91)*, CEFRIEL, Milano, Italy, 30–31 May, 1991. Italian Society of Computer Science (AICA) Press.
10. Alfonso Fuggetta and Letizia Jaccheri. Dynamic partitioning of complex process models. *Information & Software Technology*, 42(4):281–291, 2000.
11. Peter B. Henderson, editor. *Proc. 3rd ACM SIGSOFT/SIGPLAN Software Engineering Symposium on Practical Software Development Environments (Boston), 257 p.*, November 1988. In ACM SIGPLAN Notices 24(2), Feb. 1989.
12. Letizia Jaccheri. E3 project. web: http://www.idi.ntnu.no/~letizia/e3.html, April 2004.
13. Letizia Jaccheri, Mario Baldi, and Monica Divitini. Evaluating the requirements for software process modeling languages and systems. In *WORLD MULTICONFERENCE SCI/ISAS'99, International Workshop on Process support for Distributed Team-based Software Development (PDTSD'99)*, July 1999.
14. Letizia Jaccheri and Reidar Conradi. Techniques for Process Model Evolution in EPOS. *IEEE Trans. on Software Engineering*, pages 1145–1156, December 1993. (special issue on Process Model Evolution).
15. Letizia Jaccheri and Silvano Gai. Initial Requirements for E^3: an Environment for Experimenting and Evolving Software Processes. In *J.-C. Derniame (ed.): Proc. EWSPT'92, Sept. 7–8, Trondheim, Norway, Springer Verlag LNCS 635*, pages 98–101, September 1992.
16. Letizia Jaccheri, Gian Pietro Picco, and Patricia Lago. Eliciting software process models with the e3 language. *ACM Transaction on Software Engineering Methodology*, 7(4):368–410, 1998.
17. Letizia Jaccheri and Tor Stålhane. Evaluation of the E3 Process modelling language and tool for the purpose of model creation. In *PROFES 2001, 3rd International Conference on Product Focused Software Process Improvement*, Kaiserslautern, Germany, September 2001.
18. M. M. Lehman and L. A. Belady. *Program Evolution — Processes of Software Change*. Academic Press, 538 p., 1985.
19. Sun Microsystems. The Java language: A white paper. Available at http://www.java.sun.com, 1994.
20. Object Management Group. *Object Management Architecture Guide*, September 1990.
21. Richard N. Taylor, Frank C. Belz, Lori A. Clarke, Leon Osterweil, Richard W. Selby, Jack C. Wileden, Alexander L. Wolf, and Michael Young. Foundations for the Arcadia Environment Architecture. In [11], pages 1–13, November 1988.
22. Claes Wohlin, Per Runeson, Martin Hőst, Magnus C. Ohlsson, Bjőrn Regnell,

and Anders Wesslén. *Experimentation in software engineering: an introduction.* Kluwer Academic Publishers, 2000.

INDEX

Q

R

S

V

W